PERIODICALS

1. **Acta Academiae electoralis Moguntinae scientiarum utilium.** Erfordia, 1757. T. 1. Av. portrait et 7 pl. pet. in-8vo. d. veau. 12.—
Marque de bibliothèque de G. Cuvier.

2. **Acta Academiae elector. scientiarum Theodoro-Palatinae.** Pars physica. Mannheim, 1766–80. Vol. 1–4. In 6 vols. W. pl. 4to. vellum. 40.—
Since vol. III the historical and the physical sections were published separately.

3. **Acta et commentationes Imper. Universitatis Jurievensis (olim Dorpatensis).** Tartu, 1893–1917. 25 vol. Av. table des années 1893–1907. — (*Continué par*:) **Acta et commentationes** Universitatis Tartuensis (Dorpatensis). A: Mathematica, physica, medica. Vol. 1–21. B: Humaniora 1–23. C: Annales. I–X (en 4 vol.). Tartu, 1921–31. 47 vol. — Ens. 72 vol. Av. cartes, pl. et ill. 760.—
Les articles de la première série sont pour la plupart en russe; de la nouvelle série la plus grande partie est en allemand, puis en esthonien; enfin on y trouve quelques articles en français, en anglais, etc.
Collection complète jusqu'à la fin de 1931. En partie épuisé; la première série surtout est très difficile à trouver.

4. — Idem: **Acta et commentationes** Universitatis Tartuensis (Dorpatensis). A. Mathematica, physica, medica. Vol. 1–10. B: Humaniora. Vol. 1–9. Tartu, 1921–26. 19 vol. Av. cartes, pl. et ill. 90.—

5. **Acta mathematica.** Zeitschrift hrsg. von G. Mittag-Leffler u. A. Stockholm, Uppsala, 1882–1929. T. 1–52. Av. table des tom. 1–35. Ens. 53 vol. 4to. 625.—
Les contributions sont pour la plupart en allemand ou en français.

6. **Akademie van Wetenschappen, Koninklijke, te Amsterdam.** Verhandelingen, Verslagen en mededeelingen, Verslagen der zittingen der afd. Natuurkunde en Letterkunde. Jaarboek. Amst. 1854–1937. Tog. 424 vols. W. many coloured and plain pl. 4to, roy. 8vo and 8vo. Bound in ab. 300 unif. buckram bindings. *Very fine copy.*
Complete set from its commencement down to the end of 1937 of all the publications of both sections of the **Royal Academy of Sciences** at Amsterdam.
The publications are written in Dutch, English, French or German.
Fine set, uniformly bound in strong buckram bindings. Becoming extremely scarce.
Added: **Proceedings** of the Royal Academy of sciences (at Amsterdam). Amst. 1899–1937. Vols. 1–40. Tog. 54 vols. W. pl. roy. 8vo. sewed and in parts.
Extremely scarce.
See also nr. 77.

6a. **Almanach, Nieuwen,** der konstschilders, vernissers, vergulders en marmelaers voor 1777, inhoud. o.a.: Het maeken van alle sorten van water-verwen; schilderen in mignature; van het schilderen op glas; bereydinge van lakken en vernissen; verguldkonst, de wetenschap van het marmeren, enz. Bijgev. 25 levensbeschrijvingen der vermaerdste konst-schilders van Vlaenderen. Gend, (1777). 2 vols. in 1. W. 2 front., portrait and 3 pl. hfcalf.   30.—

7. **Annalen der Physik.** Hrsg. von Gilbert. Halle, 1799–1824. 76 vols. W. Index. Tog. 77 vols. unif. contempor. hfcalf.   1500.—
   A fine copy of the complete set of the precursor of Poggendorff's Annalen. Quoted in German catalogus from $ 1200 and up.

8. **Annalen van de Sterrenwacht te Leiden.** Uitgeg. d. F. Kaiser, H. G. van de Sande Bakhuysen en W. de Sitter. 's-Grav. 1868–1938. T. 1–18. 4to. cart. et br. (166.25)   105.—
   Le texte est en anglais, en allemand ou en français. En partie épuisé.
   Les tom. 1–9 ont paru sous le titre: Annalen der Sternwarte in Leiden.
   Marque de bibliothèque sur le titre des tom. 1–7.

9. **Annales de l'Observatoire de Bruxelles.** Brux. 1834–1920. Série I, 25 vol.; N. Série, t. 1–13, 14, livr. 1–3. Ens. 39 vol. 4to. 200.—

10. **Annales de l'Observatoire de Nice.** Paris, 1899–1911. T. 1–14. 14 vol. gr. in-4to.   45.—
    Les atlas des tom. 1 et 3 manquent.

11. **Annales de la Société scientifique de Bruxelles.** Brux. 1877–1928. Année 1–48. Av. table des années 1875–1901. Ens. 50 vol. Av. pl.
        260.—

12. **Annales des travaux publics de Belgique.** Brux. 1843–1932. Année 1–85. Av. tables des années 1843–1929. Ens. 88 tom. 116 vol. Av. cartes et pl. dont 24 vol. d. veau, 86 d. rel., le reste en livr. 425.—
    Très rare en état complet.

13. **Archief, uitgeg. d. h. Wiskundig Genootschap,** onder de zinspreuk: Een onvermoeide arbeid komt alles te boven, te Amsterdam. Amst. 1856–74. 3 vol. — **Nieuw Archief** voor wiskunde. Amst. 1875–93. 20 vol. — Idem. 2e Reeks. Amst. 1895–1936. T. 1–18, 19, livr. 1. Ens. 42 tom. 34 vol., dont 12 vol. d. rel., le reste en livr.
        160.—
    Publications de la Société de mathématiques à Amsterdam. Série complète.
    Voir aussi le no. 33.

14. **Archiv der Mathematik und Physik.** Hrsg. von Grunert u. A. Greifswald, etc. 1841–1919. 3 series. 70 + 17 + 28 = 115 vols. in 113. bound (not unif.).   2425.—
    All published. Extremely rare in complete state.
    All volumes in the original impression.

15. **Archives du Musée Teyler.** La Haye, 1917–38. 3e Série, t. 3–8. 6 vol. Av. pl. et figg.   14.50
    Contient e.a.: **H. A. Lorentz,** Application de la théorie des électrons aux propriétés des métaux. — **A. D. Fokker,** Inleiding tot de golvings- en de quantum-mechanica. — **Id.,** Over de akoestiek van zalen, van muziekinstrumenten en van de menschelijke stem. — **A. Schreuder,** Conondontes (Trogontherium) and castor from the Teglian Clay compared with the castoridae from other localities. — **C. J. Gorter,** Paramagnetische Eigenschaften von Salzen.
    La plupart des vol. précédents (dès le commencement en 1866) sont encore à avoir. Liste des prix de chaque livr. sur demande.

Prices are in Dutch guilders

A COLLECTION OF
BOOKS, MOSTLY OLD, ON MATHE-
MATICS, PHYSICS, ASTRONOMY,
INSTRUMENTS, MACHINES,
TECHNICAL |WORKS, INDUSTRIES,
TRADES

PRECEDED BY

A COLLECTION OF MORE THAN
TWO HUNDRED PERIODICAL SETS
AND INTERNATIONAL CONGRESSES
ON THE SAME SUBJECTS

SPRINGER-SCIENCE+BUSINESS MEDIA, B.V.

Contents:

Prices are in Dutch guilders

1 guilder is about $ 0.54

My American customers may send their orders through Messrs. TICE & LYNCH, Forwarding Agents, 21 Pearl Street, New York City.

ISBN 978-94-015-2202-1   ISBN 978-94-015-3423-9 (eBook)
DOI 10.1007/978-94-015-3423-9

Softcover reprint of the hardcover 1st edition 1939

16. **Archives néerlandaises des sciences exactes et naturelles.** Publ. par la Société Hollandaise des Sciences à Harlem. Réd. p. J. Bosscha, C. A. Pekelharing, H. Zwaardemaker, J. P. Lotsy, e.a. La Haye, Harlem, 1866–1938. Série I. 30 vol. Série II. 15 vol. Série III (A. Sciences exactes. 14 vol. B. Sciences naturelles 5 vol. C. Physiologie de l'homme et des animaux. T. 1–23). Ens. 87 vol. Av. pl. en couleurs et en noir. 575.—
Les séries IIIa et IIIb sont continuées par le périodique „Physica".

17. **Arkiv för matematik, astronomi och fysik.** Stockholm, 1922–27. Vol. 17, part 3, 4, vols. 18, 19, 20, part 1, 2. 4 vols. In parts. 15.—

18. **Atti della 1ra–8a riunione degli scienziati Italiani**, 1839–1846. Pisa, Genova, etc. 1840–47. 8 tom. 9 vol. 4to. dont 6 cart., 2 br., 1 mar. rouge. 35.—

19. **Beiblätter zu den Annalen der Physik und Chemie.** Hrsg. von J. G. Poggendorff, G. u. E. Wiedemann u. A. Lpz. 1877–1919. 43 vol. Av. table des tom. 1–30. d. veau. 175.—
Tout ce qui a paru.

20. **Beiträge zur Geschichte der Technik und Industrie.** Jahrbuch des Vereines deutscher Ingenieure. Hrsg. von C. Matschoss. Berlin, 1909–18. Année 1–8. 8 vol. Av. ill. 4to. 25.—

21. **Beobachtungen, Astronomische,** auf der Kön. Universitäts-Sternwarte in Königsberg. Königsberg, 1815–1919. T. 1–25, 27–34, 36–40, 42, 43. gr. in-4to. dont t. 1–24 en 4 vol. veau, le reste br. 40.—

22. **Berichte der Deutschen Physikalischen Gesellschaft.** Brschw. 1903–19. 17 tom. 15 vol. d. rel. (1916–1919 en livr.). 145.—
Collection complète.

23. **Bibliothèque physico-économique,** instructive et amusante. Paris, 1782—92. Année 1—11. En 18 vol. Av. pl. pet. in-8vo. veau. 36.—
Traite de l'économie rurale, des nouv. découvertes dans les arts utiles et agréables, des nouv. machines et instrumens, des médicamens nouveaux, des moyens d'arrêter les incendies, etc.
Les pp. 249–256 de la 3e année manquent.

24. **Bulletin du Groupe d'études scientifiques.** Paris, 15 juin 1910–31 déc. 1919. 92 nos. en 5 vol. chagr. bleu, fil. dor., bord. intér., tête dor., n.r. *Bel ex.* 25.—
Collection des 92 premiers numéros de cette curieuse publication dirigée par Paraf-Javal, s'occupant en partie de l'arithmétique et de la physique, puis de toutes sortes de choses, actuelles et non, renfermant e.a. le récit d'incidents politiques auxquels ont été mêlés des anarchistes, des libres-penseurs, des matérialistes et des philosophes, dont certains particulière-ment connus.

25. **Bulletin des séances de la Société française de physique.** Paris, 1905 –10. 6 vol. En livr. 35.—
Ce Bulletin est réuni depuis 1911 avec le Journal de physique.
Ajouté: **Procès-verbaux** et résumé des communications (de la Soc. franç. de physique), pendant 1911–1913, 1920–1921. Paris, 1912–14, 21, 22. 5 vol.

26. **Bulletin de la section scientifique de l'Académie Roumaine.** Bucarest, 1913–27. Year II–X, XI, part 1–3. W. index to vol. 1–5. 9 vols. W. pl. In parts. 24.—
Part 7 and foll. of year 5 are very likely not published.

27. **Bulletin de la Société internat. des électriciens.** Paris, 1898–1923. Série I, t. 15–17. Série II. 10 vol. Série III. 10 vol. Série IV, t. 1, 2, 3, livr. 1. Av. tables des séries I et II. Ens. 25 vol. Av. pl. et ill. dont 19 vol. d. rel., le reste en livr.   90.—

28. **Collection de mémoires rel. à la physique,** publ. par la Société Française de physique. Paris, 1884–91. T. 1–5. 5 vol. Av. qq. pl. 18.—
Première série complète.

29. **Communications from the Laboratory of physics at the University of Leyden** by H. Kamerlingh Onnes, W. H. Keesom, a.o. Leiden, 1885–1939. Nos. 1–253 et Suppl. 1–85. En 17 vol. d. rel., le reste en livr.   645.—
Publication très estimée, à laquelle ont contribué ou contribuent des hommes les plus compétents comme les professeurs Kamerlingh Onnes, Kuenen, Lorentz, Zeeman, Keesom e.a.
En partie épuisé.

30. **Connoissance des temps** (ou des mouvements célestes à l'usage des astronomes et des navigateurs) pour l'année 1756–1936. Publ. par le Bureau des longitudes. Paris, 1756–1934. 173 vol. Av. cartes. dont 112 vol. d. veau, le reste cart. et br.   *Vendu.*

30a. **Correspondance mathématique, Nouvelle.** Publ. p. E. Catalan et P. Mansion. Année 1874–1880. Brux. 1878—80. 6 vols. hfcalf.   100.—
All published. Extremely scarce, forerunner of Mathésis.

31. **Dingler's Polytechnisches Journal.** Augsburg, 1820–1915. T. 1–330. 330 vol. Av. tables des tom. 1–198. Ens. 336 tom. 310 vol., dont 258 d. veau unif., 52 vol. d. rel. et cart.   1100.—
Série complète, devenue extrêmement rare avec les premières années.

32. **France automobile (et aérienne), La.** Organe de l'automobilisme et des industries qui s'y rattachent. Paris, 1897–1908. Année 2–13. 12 vol. gr. in-4to. d. chagr.   150.—

33. **Genootschap, Wiskundig, onder de zinspreuk: „Een onvermoeide arbeid komt alles te boven",** Amsterdam. Amst. 1782–1932. 400.—
C o m p l e t e   s e t   f r o m   i t s   c o m m e n c e m e n t . It is composed as follows:
a. Kunst-oefeningen. 1782, 88. 2 vols. — *b.* Wiskunstige verlustiging. 1793, 95. 2 vols. — *c.* Mengelwerk van uitgelezene wis- en natuurk. verhand. 1796, 1816. 2 vols. in 1. — *d.* Wiskundig mengelwerk. 1798, 1802. 2 vols.— *e.* Wiskundige oefeningen. 1806, 09. 2 vols. — *f.* Verzameling van voorstellen. 1811, 15. 2 vols. — *g.* Wiskundige verhandelingen. 1817. (*This small part is lacking*). — *h.* Verzameling van wiskundige voorstellen. 1820–33. 6 vols. — *i.* Verzameling van nieuwe wiskundige voorstellen. 1841, 46. 2 vols.— *j.* Verslag van het verhandelde op de wintervergad. 1842–52. 1 vol. — *k.* Nieuwe wis- en natuurk. verhandelingen. 1844, 52. 2 vols. — *l.* Verzameling van wiskundige opgaven. 1846, 54. 2 vols. — *m.* Wiskundige opgaven met hare ontbindingen. 1855–74. 4 vols. — *n.* Archief. 1859–74. 3 vols. — *o.* Nieuw archief voor wiskunde. 1875–93. 20 vols. — *p.* The same. N. Series. 1894–1932. Vol. 1–17. — *q.* Wiskundige opgaven met de oplossingen. 1875–1929. Vol. 1–14.
The divisions *a—n* are bound hfcalf, the remainder in parts.
C o m p l e t e   s e t   o f   t h e   p u b l i c a t i o n s   o f   t h e   M a t h e m a t i c a l   S o c i e t y   o f   A m s t e r d a m   u p   t o   1 9 3 2 . E x t r e m e l y   s c a r c e .
See also nr. 13.

34. **Gesellschaft der Wissenschaften zu Göttingen, Kön.** Abhandlungen.

Prices are in Dutch guilders

Mathemat.-physikal. Klasse. Göttingen, 1894–1908. Série 1, t.
39, 40; série 2, t. 1–5, 6, no. 1. Ens. 8 vol. Av. cartes et pl. 4to.
50.—

35. **Gesellschaft der Wissenschaften zu Göttingen, Kön.** Nachrichten.
Mathemat.-physikal. Klasse, 1907–1924. Philolog.-histor. Klasse,
1907–1923. Berlin, 1907–25. 35 vols. W. pl.   75.—
Missing: Mathemat.-physikal. Klasse, year 1908, part 1 and titlepage
and index to year 1919. — Philolog.-histor. Klasse, titlepage and index
to year 1914.

36. **Gesellschaft der Wissenschaften, Kön. Sächsische.** Abhandlungen.
Philolog.-histor. Klasse. Bd 1–28; Mathemat.-physikal. Klasse. Bd
1–32. Lpz.1850–1911. Av. 3 vol. de tables des années 1846–1895. —
Berichte über die Verhandlungen. 2Bde; Philolog.-histor. Klasse.
Bd 1–55; Mathemat.-physikal. Klasse. Bd 1–58. Lpz. 1848–1906. —
Ens. 176 tom. 118 vol. Av. pl. d. veau (8 vol. en livr.). *Bel ex.*
4400.—

37. **Handlingar, (Kon.)** Svenska Vetenskaps-Akademiens. Stockholm,
1739–1933. 191 vols. W. index to the years 1739–1917. 2 vols. Tog.
193 vols. in 178, of which 36 calf, 41 hfcalf, the remainder boards.
1650.—
Handlingar. 1739–79. 40 vols. in 22. — Nya handlingar. 1780–1812.
33 vols. in 22. — Handlingar. 1813–54. 42 vols. in 34. — Nya följd. 1855–
1923. 63 vols. in 85. — Handlingar. 1924–33. Vol. 1–13. — Index 1739–
1917. 2 vols.
Set from the commencement, extremely scarce thus complete.

38. **Histoire de l'Académie royale des sciences** avec les mémoires de
mathématique et de physique, 1692–1751. Amst., P. de Coup,
P. Mortier, e.a. 1706–60. 90 vol. Av. tables générales 1699–1751.
4 vol. et: Eloge des académiciens. 2 tom. 1 vol. Ens. 95 vol. pet. in-
8vo. veau ancien, dos dor. *Bel ex.*   180.—
Pendant les années 1693–1698 rien n'a paru.

39. **Ingenieur, De.** Orgaan van het Kon. Instituut van ingenieurs.
's-Grav. 1886–1931. Année 1–45, 46, livr. 1–19. 45 vol. Av. tables
des années 1886–1920. Ens. 44 vol. 4to. dont 14 toile, 7 cart., le
reste en livr.   150.—
Des collections complètes de ce périodique très estimé se rencontrent
rarement. A notre collection seulement 6 nos de la 4e année et 2 titres
manquent. Le titre de l'année 1893 n'existe pas.
Voir le no. suivant.

40. **Instituut van Ingenieurs. Complete collection** from its beginning
up to the year 1930 of all the publications of the Royal Dutch
Institute of Engineers. 's-Grav. 1848–1930. 205 vols. roy. 4to.
(25 vols. 8vo). bound.   560.—
Verhandelingen. 1848–69. 20 vols. — **Uittreksels** uit vreemde tijdschriften.
1848–69. 22 vols. — **Verslagen en Notulen.** 1848–69. 21 vols. (1847 of
„Verslagen" is missing, a few pages only!). — *(Continued by:)* **Tijdschrift**
(Verhandelingen, Uittreksels, Notulen): *a.* Verhandelingen. 1869–1916.
41 vols. (all published); *b.* Notulen. 1869–1926. 56 vols. (all published);
c. Registers. 1847–1910. 5 vols. — **De Ingenieur.** 1886–1930. Vol. 1–45.
W. index to vols. 1–40.
In 1917 the „Tijdschrift" was merged with „De Ingenieur".

41. **Insurance. — Archief voor de verzekeringswetenschap** en aanverwante vakken. 's-Grav. 1895–1919. 17 vol. Av. table. — *(Continué
par:)* **Het verzekeringsarchief.** Orgaan van de Vereeniging voor de

verzekeringswetenschap. Onder red. van H. Ekama, A. O. Holwerda, H. R. Ribbius, M. van Haaften e.a. 's-Grav. 1920–38. Année 1–19. 19 vol. toile. — Ens. 36 vol. (322.50)   165.—
Revue néerlandaise de la science des actuaires.

42. **Insurance.** — **Bericht des eidgenössischen Versicherungsamtes.** Die privaten Versicherungs-Unternehmungen in der Schweiz, 1886–1921. Bern, 1888–1924. Année 1–36. 36 vol. gr. in-4to. dont 4 vol. d. rel., le reste br.   90.—

43. — **Bulletin de l'Association des actuaires belges.** Brux. 1896–1932. Nos. 1–40 (= t. 1–26). 26 tom., dont t. 1–6 en 1 vol. d. veau, le reste br.   95.—
Extrêmement rare.
Quelques titres et tables très probablement n'ont pas paru.

44. — **Mededeelingen van de Directie der Algem. Maatschappij van levensverzekering en lijfrente,** aan hare agenten. Rott. 1881–1908. Nos. 1–1394. Av. portr. En 26 vol., dont 1–21 en 4 vol. d. chagr.   60.—
Collection très intéressante d'un périodique hebdomadaire, donnant des advis sur la science d'assurance, les problèmes d'assurance, des vies d'hommes célèbres dans la science de l'assurance sur la vie, comme de Witt, C. Huygens, W. Kersseboom, J. Hudde, Lobatto, etc.
Pas dans le commerce et rarement complet.

45. — **Mitteilungen der Vereinigung schweizerischer Versicherungsmathematiker.** Bern, 1906–27. T. 1–22. 22 tom. 21 vol., dont 14 d. rel., le reste br.   90.—

46. — **Rapport du Bureau fédéral des assurances sur les entreprises privées en matière d'assurances en Suisse,** 1886–1919. Berne, 1888–1921. Année 1–34. 34 tom. 22 vol. Av. cartes et tabl. en couleurs. 4to. dont 1886–1900 en 3 vol. d. veau, le reste en livr.   85.—

47. — **Vereins-Blatt für deutsches Versicherungswesen.** Red. A. Meyer. Berlin, 1872–90. Year 1–18. 18 vols. Hfbound.   140.—
Very rare set. Titlepages to vol. 1 and 9 and index to vol. 1 are missing, probably never issued.

48. — **Veröffentlichungen des deutschen Vereins für Versicherungs-Wissenschaft.** Hrsg. von A. Manes. Berlin, 1903–14. Fasc. 1–26. 26 fasc., dont fasc. 1–17 en 4 vol. d. rel.   45.—

49. **Jornal de sciências matematicas, fisicas e naturais.** Publ. da Academia das sciências de Lisboa. Lisboa, 1918–27. T. 20–24 (= 3e Série, t. 1–5). 5 vol.   20.—

50. **Journal für Chemie und Physik.** Hrsg. von J. S. C. Schweigger. Nürnberg, Halle, 1811–19, 26–32. T. 1–27, 46–66. 48 vol. Av. pl. dont 27 vol. d. veau, 21 cart.   150.—

51. **Journal of the Franklin Institute devoted to science and the mechanic arts.** Philadelphia, 1920–28. Vol. 190–205. 16 vols. W. ill. In parts.   25.—
Vol. 190, part. 1, vol. 194, part 6 and titlepage and index and vol. 205, part. 4 are missing.
Added: **Yearbook** of the Franklin Institute. 1919–1927. Philadelphia, 1919–27. 8 vols.

52. **Journal de mathématiques élémentaires.** Publ. sous la dir. de J. Bourget et H. Vuibert. Paris, 1877–1907. T. 1–24, 27–31. 29 tom. 17 vol. 4to. cart.   65.—

Prices are in Dutch guilders

53. **Journal de mathématiques pures et appliquées.** Paris, 1856–63. Série II, t. 1–8. 8 vol. gr. in-4to. toile. 38.—
54. **Luchtvaart, De.** Geïll. tijdschrift voor luchtvaart en aanverwante vakken. Haarlem, 1910–13. Année 2–4, 5, nos. 1–13. 4 vol. Av. ill. dont 3 cart., 1 en livr. 10.—
 2 nos. manquent.
55. **Magnetism, Terrestrial.** Internat. quarterly journal. Ed. by L. A. Bauer. Chicago, 1896–1908. T. 1–13. 13 tom. 6 vol. d. rel. (t. 13 en livr.). 60.—
56. **Mathesis.** Recueil mathématique à l'usage des écoles spéciales et des établissements d'instruction moyenne. Gand, 1881–1931. T. 1–45. 45 vol., dont 5 d. veau, le reste en livr. 585.—
57. **Mémoires de la Société des sciences physiques de Lausanne,** 1783–1786. Lausanne, 1784, 89. T. 1, 2. 2 tom. 1 vol. Av. pl. 4to. d. veau. (Dos cassé). *Très rare.* 40.—
58. **Memorie della R. Accademia d. scienze dell'Istituto di Bologna.** Sezione di scienze fisiche e matematiche. Bologna, 1907–17. 6th series, vol. 4–10; 7th series, vol. 1–4. 11 vols. W. pl. sm. fol. 20.—
59. **Museum** of verzameling van stukken ter bevordering van fraaie kunsten en wetenschappen d. M. Siegenbeek. Haarlem, 1812–14. T. 1–3. cart. 10.—
60. **Nachrichten, Wochentliche,** von gelehrten Sachen a. d. J. 1740 und 1743. Regensburg, (1740, 43). 2 vol. 4to. cart. 10.—
61. **Notices of the proceedings at the meetings of the Royal Institution of Great Britain.** London, 1906–27. Vol. 17–24, 25, part 1, 2. 7 vols. W. pl. and ill. In parts. 40.—
 Vol. XVII, pt 1 is missing.
62. **Observations de Poulkova** (*plus tard*: Publications Observatoire Central Nicolas). St. Pétersbourg, 1869–1911. Série I, t. 1–14, av. 3 suppl.; Séric II, t. 1–3, 5–19. Ens. 35 vol. pet. in-fol. cart. et br. 150.—
 T. 4 de la 2e série probablement n'a pas paru.
63. **Observations made at the Royal magnetical and meteorological observatory at Batavia.** Publ. by J. Boerema, 1924–1930. Batavia, 1928–39. T. 47–58 A et C. 12 vol. fol. 120.—
64. **Observations made at Secondary [magnetical and meteorological] stations in the Netherlands Indies.** Publ. by S. W. Visser a.o., 1919–1931. Batavia, 1930–39. T. 8A–15, 17. 8 vol. gr. in-4to. 55.—
65. **Photography, Cinematography.** — **Agfa-Photoblätter.** Berlin, 1924–38. Year 1–15. W. numer. pl. and ill. In 15 orig. portfol. 50.—
66. — **Bulletin (de l') Association belge de photographie.** Brux. 1874–1933. 55 années. — (*Continué par:*) **Bulletin** (de l') **Association belge de photographie et de cinématographie.** Brux. 1934–36. Année 1–3. — Ens. 58 tom. 57 vol., dont 25 d. veau, le reste en livr. 325.—
67. — **Camera craft.** Photographic monthly. San Francisco, 1900–38. Vol. 1–45. 45 vols. in 39. W. numer. pl. and ill. bound. 350.—

68. **Photography, Cinematography.** — **Ciné-Miroir.** Paris, 1er mai 1922—10 févr. 1939 (= Nos. 1 à 723). 4to. En nos. 75.—
Collection bien complète de cette publication cinématographique profusément illustrée.
Coupure à la première page des nos. 71 et 72.

69. — **Focus.** Fotoblad voor Groot-Nederland. Bloemendaal, 1914–32. Year 1–19. 19 vols. W. numer. pl. and ill. 4to. orig. cloth and hfcloth. 165.—
One of the most important Dutch periodicals on photography.

70. — **Lux.** Tijdschrift tot bevordering der fotografie en aanverwante kunsten en wetenschappen. Amst. 1889–1933. 44 years in 44 vols. W. numer. pl. and ill. 4to and 8vo. bound. 300.—
The years 1928–1933 are entitled: Vereenigde Fotobladen Lux-De Camera.
All published. Very scarce.

71. — **Rolprent, De.** Hollandsch weekblad voor de moderne film. Amst. 1925–27. 3 years (= 69 nos.). W. numer. portr., pl. and ill. large 4to. Bound in 1 stout vol., illustr. covers preserved. 20.—
The leading Dutch periodical on modern films. All published.

72. — **Tout,** hebdomadaire du reportage photographique. Anvers, 1932, 33. Année 1, 2. 2 vol. Av. des milliers d'ill. fol. d. rel. 20.—

73. **Physica.** Nederlandsch tijdschrift voor natuurkunde. Red. A. D. Fokker, E. Oosterhuis, B. van der Pol. Eindhoven, 1921–33. 13 vol. Av. table des années I–X. Ens. 14 vol. Av. pl. et figg. toile. (125.—) 60.—
Collection complète. Parmi les collaborateurs citons: H. C. Burger, les Prof. P. Ehrenfest, W. J. de Haas, W. H. Keesom, H. A. Lorentz, H. Kamerlingh Onnes, J. D. van der Waals, P. Zeeman, e.a.

74. — Idem. 14 vol. En livr. (100.—) 40.—

75. **Physica.** Onder red. van J. D. van der Waals Jr., P. Zeeman, A. D. Fokker, W. J. de Haas, W. H. Keesom, L. S. Ornstein e.a. 's-Grav. 1934–38. T. I–V. 5 vol. Av. figg. toile. (Forts vol.). 137.50
Ce périodique se compose des études originales qu'écrivent les physiciens néerlandais. Les articles sont' en allemand, en anglais ou en français. Il forme une continuation aux séries III A et III B (Sciences exactes et Sciences naturelles) des Archives, publ. par la Société Hollandaise des sciences à Harlem. (Voir le no. 16).

76. **Proceedings of the American Academy of arts and sciences.** Boston, 1913–27. Vol. 48–61, 62, part 1–8. 14 vols. W. pl. In parts. 20.—
Vol. 49, part 1 and 2, vol. 50, part 6 and vol. 58, part 8 are missing.

78. **Proceedings of the Indian Association for the cultivation of science.** Calcutta, 1920–26. Vol. 6–9, 10, part 1. 4 vols. W. pl. In parts.
20.—
Added: **Report** of idem, 1918. Calcutta, 1920.

79. **Radio-nieuws.** Orgaan van de Nederlandsche vereeniging voor radio-telegrafie. Red. J. Corver. 's-Grav. 1925–30. Année 8, no. 6 à fin, 9–13. 6 vol. Av. ill. En livr. (± 50.—) 15.—
Les titres et tables n'existent pas.

80. **Rendiconto** d. sessioni della R. Accademia d. scienze d. Istituto di Bologna. Bologna, 1907–17. New Series, vol. 11–21. 11 vols. 18.—

81. **Revue de l'aéronautique théorique et appliquée.** Dir. H. Hervé. Paris, 1888–92. Year 1–5. 5 vols. W. ill. roy. 4to. In parts. 25.—
One titlepage and 10 pl. are missing.

Prices are in Dutch guilders

82. **Revue de mathématiques spéciales.** Réd. par B. Niewenglowski.
Paris, 1890–1914. Année 1–13, 14, livr. 1–9. 4to. dont t. 1–20 en
10 vol. cart. 70.—
Série complète. Cette revue ne fut continuée qu'en 1920.

83. **Revue des questions scientifiques** publ. par la Société scientifique
de Bruxelles. Louvain, 1877–1931. T. 1–100 (= 4e série, t. 20).
Av. table des tom. 1–80. Ens. 101 vol. Av. cartes, pl. et ill. dont
75 vol. d. rel., le reste en livr. 500.—
Revue générale des sciences exactes et naturelles.

84. **Revue scientifique, La,** de la France et de l'étranger (plus tard:)
Revue scientifique (Revue rose). Paris, 1867–1909. T. 4–6, 11–84.
41 vol. Av. ill. 4to. dont 38 vol. cart., 3 d. rel. 60.—
Les premières années sous le titre: Revue des cours scientifiques.
Marque de bibliothèque sur les titres. Manquent les pp. 789–812 du
t. 21, les pp. 640–673 du t. 31, les pp. 289–320 du t. 61 et 12 titres.

85. **Revue semestrielle des publications mathématiques.** Réd. par P. H.
Schoute, D. J. Korteweg, e.a. Amst. 1893–1934. 39 vol. Av. table
des matières des tom. 1–30. 220.—
Tout ce qui a paru.

86. **Schauplatz** der Natur und der Künste in vier Sprachen, deutsch,
latein., französ., und italien. Wien, 1774–83. Année 1–10. 5 vol.
Av. 480 pl. p. E. Möszmer, Schmutzer, I. Wagner, e.a. 4to. vél.
Bel ex. 90.—
Les 480 pl. représentent e.a. toutes sortes de métiers, comme: le fondeur
de caractères, l'imprimeur en taille-douce, le graveur en bois, l'ouvrier en
soie, le cordier, le tonnelier, le parfumeur, le tanneur, le boutonnier, le gan-
tier, le pêcheur des perles, l'oiseleur, etc.

87. **Scripta Universitatis atque bibliothecae Hierosolymitanarum.** Auct.
Concilii Acad. A. Besredka, A. Einstein, J. Hadamard, E. Landau,
e.a. ed. I. Velikovsky et H. Loewe. Mathematica et physica. Vol.
I. Cur. A. Einstein. Hierosol. 1923. 4to. 10.—

88. **Skrifter Videnskaps-Selskabet (Videnskaps-Akademi) i Christiania.**
Matematisk-naturvidenskapelig Klasse. Oslo, 1894–1925. 40 vol.
W. pl. 245.—
Complete set from its beginning.

89. **Taschenbuch der Luftflotten.** München, 1925. Année 4. Av. de
nombr. ill. et figg. pet. in-8vo. toile. (7.20) 3.50

90. **Stahl und Eisen.** Zeitschrift für das deutsche Eisenhüttenwesen.
Hrsg. von Beumer und Petersen. Düsseldorf, 1900, 01, 11–21.
Année 20, 21, 31–41. 24 vol. Av. pl. et figg. 4to. dont 11 vol. d.
veau, 9 d. rel., le reste en livr. 100.—
Manquent année 21, livr. 23, année 41, pl. 3 et 4 de la 1re livr.

91. **Technique moderne, La.** Revue mensuelle illustrée des sciences
appliquées à l'industrie, au commerce et à l'agriculture. Réd. G.
Bourrey. Paris, 1909–21. T. 1–13. 13 vol. Av. Suppl. des tom.
1–6. 6 vol. Ens. 19 vol. Av. pl. et ill. gr. in-4to. dont 5 vol. d. veau,
le reste br. et en livr. 80.—

92. **Tijdschrift voor reken-, stel- en meetkunst,** met aanh. over de be-
schouwende gedeelten der wiskunst, de natuurkunde, enz. Dordrecht, 1839–49. 9 vol. Av. pl. de figg. pet. in-8vo. cart. orig. 15.—
Collection complète.

93. **Tijdschrift, Nederlandsch,** voor natuurkunde. Red. M. Minnaert, E. Oosterhuis, B. van der Pol, C. Zwikker. 's-Grav. 1934–38. Année 1–5. 5 vol. Av. figg. toile. 45.—

Ce périodique est la continuation directe de l'ancien périodique „Physica" (voir le no. 73). Les articles sont en néerlandais et contiennent des aperçus synthétiques sur des questions de physique ayant un caractère d'actualité, des comptes rendus de travaux scientifiques originaux publiés en Hollande, des notices analytiques et critiques sur des livres parus, des rapports sur des conférences et des controverses, des questions rel. à l'enseignement etc. Parmi les collaborateurs citons e.a. les prof. Clay, Dorgelo, Fokker, de Haas, Keesom, Kramers, van der Waals, Zeeman e.a.

95. **Tijdschrift, Wiskundig.** Onder red. van Vaes, Krediet en Quint. Haarlem, 1905–21. En livr. 50.—

Tout ce qui a paru.

96. **Verhandelingen van het Bataafsch Genootschap der proefondervindelijke wijsbegeerte te Rotterdam.** Rott. 1774–98. 12 vols. — Nieuwe verhandelingen. Rott. 1800–65. 12 vols. — Nieuwe verhandelingen. 2e reeks. Rott. 1867–1925. Vols. 1–9. — Tog. 33 vols. in 26. W. pl. 4to. of which 16 vols. hfcalf, the remainder sewed. 200.—

Batavian Society of natural sciences of Rotterdam. All published till 1925. Very rare.
Added: Verslag der voordrachten van leden van het Bataafsch genootschap. Rott. 1914–26. Bundel 1, 2, 3, livr. 1. 2 vols. W. pl. In parts.

97. **Vriend der wiskunde, De.** Tijdschrift voor allen, die examen in dat vak moeten afleggen. Red. A. J. van Breem. Schoonhoven, 1886–1916. Année 1–31. — Supplement. Schoonhoven, 1887–1913. 25 vol. — Ens. 56 vol. cart. 80.—

Ajoutée une table des matières en ms. par J. G. van Deventer.

98. **Wiadomosci matematyczne.** Red. S. Dickstein. Warszawa, 1905–14. Vol. 9—17, 18, part 1–4. 9 vols. In parts. per vol. 3.50

99. **Zeitschrift für Instrumentenkunde.** Organ für Mittheilungen a. d. ges. Gebiete der wissenschaftl. Technik. Mit Beiblatt. Berlin, 1881–1930. Année 1–50. Av. tables génér. des années 1–50. 81 tom. 53 vol. Av. portr., pl. et ill. gr. in-8vo et 4to. dont 15 vol. d. veau, le reste d. rel. 1575.—

Collection depuis l'origine, extrêmement rare.

100. **Zeitschrift für Mathematik und Physik.** Hrsg. von O. Schlömilch und B. Witzschel. Lpz. 1856–62. Année 1–7. Av. Literaturzeitung année 1–5. Ens. 12 tom. 7 vol. d. veau. 150.—

Les titres et tables de la 2e année manquent; 4 pl. en facs.

101. **Zeitschrift für Physik.** Hrsg. von K. Scheel. Berlin, 1929–32. T. 57–74. 18 vol. Av. pl., figg. et ill. En livr. 100.—

102. **Zeitschrift, Elektrotechnische.** (Centralblatt für Elektrotechnik). Red. von E. Zetzsche, R. Ruhlmann, u. A. Berlin, 1880–1927. Année 1–47, 48, nos. 1–25. 50 vol. Av. figg. 4to et fol. 25 vol. d. veau, 16 d. rel., 6 cart., 1 en livr. 145.—

103. **Zeitschrift, Physikalische.** Hrsg. von E. Riecke, H. Th. Simon u. A. Lpz. 1900–32. Year 1–32, 33, nos. 1–6. 32 vols. W. pl. 27 vols. unif. hfcloth, 5 vols. in parts. 540.—

Prices are in Dutch guilders

INTERNATIONAL CONGRESSES

104. **Congrès internat.**, 3e, de l'acétylène. Paris 1900. Rapports, discussions, travaux et résolutions. 7.50

105. **Congress, 12th Internat.**, of acetylene, oxy-acetylene welding and allied industries. London 1936. Proceedings. 6 vols. W. numer. portr., ill. and figg. fol. 27.50

106. — **2d Internat. actuarial.** London 1898. Transactions. toile. 7.50

107. — **5th Internat.**, of actuaries. Berlin 1906. Reports, memoirs and proceedings. cloth. 18.—

108. **Congrès internat.** d'aéronautique. Paris 1889. Procès-verbaux sommaires. 1.—

109. — 3e, d'aéronautique. Milan 1906. Rapports et mémoires. 6.—

110. [**Conférence**], 3e à 6e, de la Commission internat. pour l'aérostation scientifique. Berlin 1902, St. Pétersbourg 1904, Milan 1906, Monaco 1909. Procès-verbaux des séances et Mémoires. 4 vols. *Rare.*
Each vol. 3.50

111. **Congrès internat.** des applications de l'éclairage. Paris 1937. Comptes rendus. W. figg. 4to. 16.50

112. — des applications de l'électricité. Marseille 1908. Rapports prélimin. Organisation. 3 vols. W. figg. 15.—

113. — des applications électrocalorifiques et électrochimiques. Schéveningue 1936. Recueil des travaux et compte-rendu des séances. W. figg. bound. 6.—

114. — des applications du moteur à mélange tonnant et du moteur à combustion interne aux marines de guerre, de commerce, de pêche et de plaisance. Paris 1908. Rapports. 8.—

115. **Congress, 5th Internat.**, for applied mechanics. Cambridge (Mass.) 1938. Proceedings. W. figg. 4to. bound. 15.—

116. **Congrès internat.** d'automobilisme. Paris 1900. [Rapports]. 6.—

117. — 2e, d'automobilisme. [Rapports]. Paris 1903. 2 vols. 12.50

118. — 1er, du béton et du béton armé. Liége 1930. [Rapports]. (2e éd.). 2 vols. W. pl. and ab. 1200 figg. roy. 4to. 32.—

119. — des brasseurs. Paris 1878. Comptes rendus sténograph. 5.—

120. — 3e, des chemins de fer. Paris 1889. Compte rendu général. 3 vols. 4to. bound. 22.—

121. — 4e, des chemins de fer. St. Pétersbourg 1892. Compte rendu général. 4 vols. 4to. 25.—

122. — 5e, des chemins de fer. Londres 1895. Compte rendu général. 4 vols. 4to. 24.—
Added: Compte rendu sommaire.

123. — 6e, des chemins de fer. Paris 1900. Compte rendu général. 6 vols. 4to. 36.—

124. — 7e, des chemins de fer. Washington 1905. Compte rendu général. 3 vols. W. pl. 4to. bound. 15.—

125. — 8e, des chemins de fer. Berne 1910. Compte rendu. 2 vols. 4to. bound. 24.—

12   INTERNATIONAL CONGRESSES

126. **Congrès,** 13e ,(de l')Association internat. des chemins de fer. Paris
1937. Rapports. 4to. In parts.   17.50
    Publ. also in Bulletin de l'Association internat. du Congrès des chemins
    de fer.
127. **Congrès internat.** de chronométrie. Paris 1889. Comptes rendus, procès-verbaux, rapports et mémoires. 4to.   5.—
128. — de la construction métallique (steel construction). Liége 1930.
Comptes rendus des séances techniques. Conclusions générales.
4to.   4.—
129. **Congres, Internat. electrical.** St. Louis 1904. Transactions. 3 vols. W.
pl. and figg.   15.—
130. **Congrès internat.** des électriciens. Paris 1881. [Compte-rendu sténograph.] des séances. bound.   7.50
131. — des électriciens. Paris 1889. Comptes rendus des travaux.   6.—
132. — d'électricité. Paris 1900. Rapports et procès-verbaux. Annexes.
2 vols. W. figg.   7.50
133. **Congress, Internat. engineering.** Chicago 1893. Papers read before
division A. Civil engineering. 2 vols. W. numer. maps and pl. bound.
                                                              5.—
    Transactions of the Amer. Society of Civil Engineers. Vol. 29 and 30.
134. **Congrès internat.** pour l'essai des matériaux. Amsterdam 1927. 2
vols. W. numer. figg. etc. bound. (30.—)   15.—
135. **Congrès** (de l')Association internat. pour l'essai des matériaux.
Zürich 1931. 2 vols. W. figg. bound.   35.—
136. **Congrès internat.** des méthodes d'essai des matériaux de construction. Paris 1900. Communications. Documents particuliers.
Rapports généraux. Procès-verbaux in extenso. 6 vols. in 7. W. pl.
and figg. roy. 4to.   24.—
137. **Convegno** di fisica nucleare. 1931. W. 1 pl. and figg.   3.50
    R. Accad. d'Italia. Fondaz. A. Volta. Atti dei convegno. I.
138. **Congrès internat.** de fonderie. Paris 1937. Mémoires.   12.—
139. — 1er, du froid (refrigeration). Paris 1908. Comptes rendus.
Rapports et communications. 3 vols.   15.—
140. — 6e, du froid. Buenos Aires 1932. Actes. 7 vols. W. figg.   40.—
141. — 7e, du froid. La Haye 1936. 4 vols. W. figg.   25.—
142. — de génie civil. Paris 1878. Comptes rendus sténograph.   6.—
143. **Conférence générale,** 12e, de l'Association géodésique internat.
Stuttgart 1898. Comptes-rendus. W. 38 maps and pl. 4to.   5.—
144. **Congrès internat.,** (1er), des géomètres-experts. Paris 1878. Comptes
rendus sténograph. *Rare.*   12.—
145. — 2e, des géomètres. Bruxelles 1910. Compte-rendu. bound. 5.—
146. **Conférence internat.** du goudron pour route. Scheveningen 1938.
Rapport: Les propriétés du goudron pour routes intéressantes du
point de vue économique et technique par H. Mallison.   1.—
147. — 6e, des grands réseaux électriques à haute tension. Paris 1931.
Compte rendu des travaux. 3 vols. W. figg. bound.   25.—
148. — 8e, des grands réseaux électriques à haute tension. Paris 1935.
Comptes rendus des travaux. 3 vols. W. figg. bound.   30.—
149. — 9e, des grands réseaux électriques à haute tension. Paris 1937.
Compte rendu. 3 vols. bound.   30.—

Prices are in Dutch guilders

150. **Conférence internat.** de l'heure. Paris 1912. [Procès-verbaux des séances. Annexes]. 4to. 7.50
151. **Meeting, 1st,** (of the) Internat. Association for hydraulic structures research. Berlin 1937. Report. bound. 12.50
152. **Congrès internat.,** 3e, des associations d'inventeurs et d'artistes industriels. Bruxelles 1910. Actes. 3.—
153. — (Pays Alliés et Neutres) des Associations d'inventeurs, d'artistes industriels etc. Bruxelles 1919. [Compte rendu. Rapports]. 4to. *Rare.* 5.—
154. **Kongress, 2er Internat. Kälte-.** Wien 1910. Bericht. 2 vols. W. figg. 15.—
155. **Congress, 2d (Internat.),** on large dams. Washington 1936. Transactions. 5 vols. bound. 60.—
156. **Congresso internaz.,** 4°, dei matematici. Roma 1908. Atti. 3 vols. 15.—
157. — dei matematici. Bologna 1928. Atti. 6 vols. bound. 35.—
158. **Kongress, 2e Skandinaviske Matematiker.** Kjøbenhavn 1911. Beretning. cloth. 4.—
159. — 3e **Skandinaviske Matematiker-.** Kristiania 1913. Beretning. 4.50
160. — 6e **Skandinaviske Matematiker-.** København 1925. Beretning. 8.25
161. — 7e **Skandinaviske Matematiker-.** Oslo 1929. Comptes rendus. 6.50
162. **Kongressen, 8e Skandinaviska Matematiker-.** Stockholm 1934. Compte rendu. 12.—
163. **Congress, Internat. mathematical.** Toronto 1924. Proceedings. 2 vols. roy. 4to. bound. (1941 pp.). 30.—
164. **Congrès internat.,** 2e, des mathématiciens. Paris 1900. Compte rendu, Procès-verbaux et communications. 8.—
165. — des mathématiciens. Strasbourg 1920. Comptes rendus. gr. in-4to. 16.—
166. — des mathématiciens. Oslo 1936. Procès-verbaux et conférences générales. Comptes rendus. 2 vols. 15.—
167. **Kongress, 3er Internat. Mathematiker-.** Heidelberg 1904. Verhandlungen. 7.50
168. **Kongress, 5er,** der Skandinavischen Mathematiker. Helsingfors 1922. Wissenschaftliche Vorträge. W. pl. 5.—
169. **Congrès internat.** de météorologie. Paris 1878. Comptes rendus sténograph. 6.—
170. — de météorologie. Paris 1900. Procès-verbaux des séances et Mémoires. 7.50
171. **Conférence internat.** de navigation aérienne. Paris 1910. Procès-verbaux des séances et annexes. fol. 12.50
172. — 5e, de la navigation aérienne. La Haye 1930. 2 vols. W. ab. 500 ill. and figg. (LXVIII and 1741 pp.). 10.—
173. — The same work. 2 vols. bound. (36.—) 14.—
174. **Congrès internat.** du pétrole. Bucarest 1907. Compte rendu. II. Mémoires. W. maps and figg. (913 pp.). 17.50
Ce volume est le plus important: le t. I contient les Préparatifs et marche du Congrès et les débats par sections et ne compte que 381 pp.

175. **Congrès mondial,** 2e, du pétrole. Paris 1937. Comptes rendus. 5
     vols.   70.—
176. — The same work. 5 vols. bound.   77.50
177. **Congrès internat.,** 5e, de photographie. Bruxelles 1910. Compte-
     rendu, procès-verbaux, rapports, notes et documents. *Out of print.*
                                                                    7.50
178. — de physique. Paris 1900. Rapports. Travaux. 4 vols.   16.—
179. **Conference, 1st World Power.** London 1924. Transactions. 5 vols.
     W. numer. maps, ill. and figg. bound. ($\pm$ 7000 pp.).   75.—
180. **Congrès,** 6e, (de l')Union internat. des producteurs et distributeurs
     d'énergie électrique. Schéveningue 1936. Compte rendu. Rapports.
     3 vols. fol.   35.—
181. — 5e, des ingénieurs en chef des associations de propriétaires
     d'appareils à vapeur. Lyon 1880. Compte rendu des séances. 2.—
182. — 9e, des ingénieurs en chef des associations de propriétaires
     d'appareils à vapeur. Paris 1884. Compte rendu des séances. 2.—
183. — 10e, des ingénieurs en chef des associations de propriétaires
     d'appareils à vapeur. Paris 1885. Compte rendu des séances. 2.—
184. **Conférence européenne** des radiocommunications. Le Caire 1938.
     Documents. 2 vols. 4to.   24.—
185. **Conférence internat.** de radiocommunications. Londres 1912. Docu-
     ments. 4to.   6.—
186. **Réunion,** 4e, du Comité consultatif internat. technique des radio-
     communications. Bucarest 1937. Documents. 2 vols. 4to.   20.—
187. **Congrès internat.** de radiologie et d'électricité. Bruxelles 1910.
     Comptes rendus. 2 vols. W. pl. and figg.   12.—
     I. Sciences physiques. — II. Sciences biologiques. Radiologie médicale.
188. — 1er, pour l'étude de la radiologie et de l'ionisation. Liége 1905.
     Comptes rendus.   9.50
189. **Conférence radiotélégraphique internat.** Washington 1927. Propo-
     sitions. 4to.   5.—
190. **Conference, Internat. road tar.** Scheveningen 1938. Paper: Road tar
     properties of economic and technical importance by H. Mallison.
                                                                     1.—
191. **Congrès internat.,** 1er, de la route. Paris 1908. Compte rendu des
     travaux. W. pl. bound.   5.—
192. — 2e, de la route. Bruxelles 1910. Compte rendu des travaux.
     Rapports généraux. Rapports et communications. 1 vol. et 122 fasc.
     toile et en livr.   15.—
193. — 3e, de la route. Londres 1913. Compte rendu des travaux. W.
     portr. and pl. bound.   7.50
194. — 4e, de la route. Seville 1923. Compte rendu des travaux. Rap-
     ports. 1 vol. et 53 fasc. Ens. 2 vol. toile et en portef.   12.—
     Les nos. 26, 30, 41, 50, 52, 57 des Rapports n'ont pas paru.
195. — 5e, de la route. Milan 1926. Rapports. 48 fasc. En portef. 10.—
     Les nos. 1, 19, 20, 27, 36, 47, 49 n'ont jamais paru.
196. — 7e, de la route. München 1934. Compte rendu des travaux. 6.50
     To be had also with english text (:Proceedings (of the) 7th Internat.
     road congress) or with german text (:Bericht über die Arbeiten (des) 7.
     Internat. Strassenkongresses).

Prices are in Dutch guilders

197. **Congrès internat.**, 8e, (de l') Association Internat. Permanente de la route. La Haye 1938. Rapports. En portef. 20.—
198. — 1er, de la sécurité aérienne. Paris 1930. 5 vols. 4to. 15.—
199. **Kongress, 22er Internat.**, der Strassenbahnen, Kleinbahnen und der öffentl. Kraftfahrunternehmen. Warschau 1930. Ausführl. Bericht. 4to. 25.—
200. — **23er Internat.**, der Strassenbahnen, Kleinbahnen und der öffentl. Kraftfahrunternehmen. Haag 1932. Ausführl. Bericht. 27.50
201. **Konferenz, Internat. Straszenteer.** Scheveningen 1938. Bericht: Die wirtschaftlich und technisch wichtigen Eigenschaften des Straszenteers von H. Mallison. 1.—
202. **Congrès internat.** de surveillance et de sécurité en matière d'appareils à vapeurs. Paris 1900. [Rapports. Procès-verbaux]. W. 2 pl. 7.50
203. **Conférence télégraphique et téléphonique internat.** Le Caire 1938. Documents. 2 vols. 4to. 17.50
204. **Assemblée générale**, 9e, (de l')Union Internat. Permanente de tramways. Stockholm 1896. Procès-verbal. fol. 10.—
205. **Congrès internat.**, 19e, de tramways, de chemins de fer d'intérêt local et de transports publics automobiles. Paris 1924. Comptes rendus. 4to. W. maps. 15.—
206. — 20e, de tramways, de chemins de fer d'intérêt local et de transports publics d'automobiles. Barcelone 1926. Comptes-rendus. 4to. 18.—
207. — 23e, de tramways, de chemins de fer d'intérêt local et de transports publics d'automobiles. La Haye 1932. Comptes-rendus. 4to. 27.50
208. — 24e, de tramways, de chemins de fer d'intérêt local et de transports publics d'automobiles. [Rapports]. Berlin 1934. 4to. 30.—
209. — 25e, de tramways, de chemins de fer d'intérêt local et de transports publics automobiles. Vienne 1937. Comptes rendus. [Rapports]. 4to. 55.—
210. — de l'utilisation des eaux fluviales. Paris 1889. Compte-rendu des travaux. W. map, pl. and ill. bound. 5.—
211. **Kongress, 6er Internat.**, für Versicherungs-Wissenschaft. Wien 1909 Gutachten. Denkschriften und Verhandlungen. 3 vols. in 4. bound. 20.—

National Congresses

212. **Congrès français**, 2e, du froid. Toulouse 1912. Comptes rendus, rapports et communications. 2 vols. bound. 10.—
213. **Congrès** de la houille blanche. Grenoble, etc. 1902. Compte rendu des travaux, des visites industrielles et des excursions. 2 vols. W. numer. maps, pl. and ill. bound. 15.—
214. — 16e, (de la) Société technique de l'industrie du gaz en France. Paris 1889. Compte rendu. W. 29 pl. bound. 7.50
215. **Congreso nacional**, 1er, de ingenieria. Madrid 1919. Trabajos. 4 vols. (2515 pp.). 20.—

216. **Congrès** des ingénieurs électriciens. Paris 1913.   5.—
217. — (du) Syndicat professionnel des producteurs et distributeurs d'énergie électrique. Grenoble 1925. [Rapports et discussions]. 4to.
6.—

MATHEMATICS

218. **Arfe y Villafañe, J. de,** Varia comensuracion. Ed. aument. p. J. Assensio y Torres. Madrid, 1806. 2 vols. in 1. W. 103 pl. fol. calf. 24.—
1. D e g e o m e t r i a; 2. Gnomónica; 3. Arquitectura; 4. De medidas del cuerpo humano; 5. De las dimensiones, coloridos, y algunas propiedades de los animales quadrupedos. 6. De las dimensiones etc. de las aves; 7. De las piezas de plata y oro destinadas en las iglesias para el servicio del culto divino; 8. Apéndice en el qual se trata de la ciencia heraldica o del blason.

219. **Arnaiz, M.,** El espiritu matematico de la filosofia moderna. (C.) Contestacion p. J. Zaragüeta. (Madrid), 1923.   3.50

220. **Bachmann, P.,** Das Fermatproblem in seiner bisherigen Entwicklung. Berlin, 1919.   1.25

221. **Bartjens, W.,** De vernieuwde cijfferinge geheel uitgewerkt. Met toegift en een korte practijk om alle specien en geld tot guldens te maken door H. Haanstra. Waarby gevoegd de Rekenkonst.... d. S. Nieuwenhuis. 4e dr. Leeuwarden, A. Ferwerda, 1752. 2 tom. 1 vol. 4to. d. veau. (Taché).   10.—

222. **Bentley, A. F.,** Linguistic analysis of mathematics. Bloomington Ind. 1932. toile.   6.—

223. **Bézout,** Suite du cours de mathématiques, à l'usage des gardes du pavillon et de la marine. Paris, 1765–66. 5 vol. Av. pl. veau ancien, dor. s. tr. *Bel ex.*   25.—
I. Le traité de navigation. Av. 10 pl. — II. Eléments de géométrie et la trigonométrie. Av. 7 pl. — III. L'algèbre. Av. 4 pl. — IV, V. Principes généraux de la méchanique. 2 vol. Av. 16 pl.

224. **Biermann, O.,** Theorie der analyt. Functionen. Lpz. 1887. dos et coins veau. *Epuisé.*   6.—

225. **Bliedner, E.,** Philosophie der Mathematik bei Fries. Jena, 1904. — **L. Koenigsberger,** Zur Erinnerung an Jacob Friedrich Fries. Heidelberg, 1911. — **K. M. Poeschmann,** Das Wertproblem bei Fries. Altenburg, 1905.   6.—

226. **Boole, G.,** Treatise on differential equations. 4th ed. W. suppl. vol. London, 1877, 65. 2 vol. toile. *Rare.*   8.50

227. **Borghi, P.,** Libro de Abacho. — Incomincia la nobile opera de arithmetica ne laqual se tratta tutte cose a mercantia pertinente. Vinegia, Franc. Bindoni et M. Pasini, 1560. 4to. *Rare.*   15.—
Les pp. 7 et 8 manquent. Sur le titre la date 1540.

228. **Bowley, A. L.,** The mathematical groundwork of economics. Oxford, 1924. toile.   2.—

229. **Bramer, B.,** Apollonius Cattus, oder geometrischer Wegweiser. Cassel, 1646, 47. 2 tom. 1 vol. Av. 30 pl. se dépliant et ill. dans le texte. 4to. vél.   24.—

230. **Briot** et **Bouquet,** Théorie des fonctions elliptiques. 2e éd. Paris, 1875. 4to. d. veau. *Epuisé.*   16.—

Prices are in Dutch guilders

231. **Bruggen, J. van,** De hoofdstelling der axonometrie. Utrecht, 1915.
2.—

232. **Cantor, M.,** Vorlesungen über Geschichte der Mathematik. Von den ältesten Zeiten bis 1799. Lpz. 1880–1908. 4 vol. d. veau. 75.—
Tous les vol. en impression originale. (La réimpression anastatique a mal réussi).

233. **Ceulen, L. van,** Vanden circkel, daer in gheleert werdt te vinden de naeste proportie des circkels-diameter teghen synen omloop. Noch de tafelen Sinuum...., hoog-noodigh voor de land- meters .... Ten laetsten van interest, met allerhande tafelen. 2e ed. uyt-geg. d. S. v. d. Eycke. Leyden, I. Azn. v. d. Marsse voor J. v. Colster, 1615. Av. figg. 4to. vél. 20.—

234. **Chernac, L.,** Cribrum arithmeticum sive tabula continens numeros primos, a compositis segregatos, occurrentes in serie numerorum ab unitate progredientium usque ad 1.020.000 etc. Daventria, 1811. Stout vol. 4to. hfcalf. 25.—

235. **Cohen, H.,** Platons Ideenlehre und die Mathematik. Marburg, 1878. 4to. *T. à p.* 9.—

236. **Colangelo, F.,** Storia dei filosofi dei matematici Napolitani, e delle loro dottrine da Pitagorici sino al secolo XVII dell'era volgare. Napoli, 1833–34. 3 tom. 1 vol. 4to. cart. 28.—
Ouvrage estimé.
La marge infér. au commencement avec une tache d'eau.

237. **Comte, A.,** Traité élémentaire de géometrie analytique, à deux et à trois dimensions. Paris, 1843. d. veau. 10.—

238. **Coutereels, J.,** Arithmetica, nieuwelycx oversien d. C. Fr. Eversdyc. Middelburg, 1658. pet. in-8vo. d. veau. 7.50

239. — 't Konstigh cijfferboek, met uytwerkingh aller questien d. D. de Grauw. Utrecht, 1690. Av. joli front. p. J. Luyken. pet. in-8vo. vél. 7.50

240. **Dassen, C. E.,** Etude sur les quantités mathématiques grandeurs dirigées-quaternions. Paris, 1903. 1.50

241. **Davis, H. T., a.o.,** Tables of the higher mathematical functions. Bloomington, Ind. 1933. toile. 11.50

242. **Dechales, C. L. Milliet,** Cursus seu mundus mathematicus, complectens Euclidis ll. VIII, arithmeticam, Theodosii sphaerica, trigonometriam, geometriam pract., mechanicam, staticam, geographiam univers., tractatum de magnete, architecturam civilem, et artem tignariam. Lugd. 1674. 3 vol. Av. figg. fol. veau. 25.—

243. — Même ouvrage. Ed. 2a stud. A. Varcin. Lugd. 1690. 4 vol. Av. figg. fol. vél. 35.—

244. **Diophantus Alexandrinus,** Arithmeticorum ll. VI, et de numeris multangulis liber unus (Gr. et lat.). Cum commentt. C. G. Bacheti et observatt. P. de Fermat. Acc. doctrinae analyticae inventum novum, collectum ex variis ejusdem de Fermat epistolis. Tolosa, 1670. W. figg. fol. vellum. 120.—
Valuable edition, containing for the first time the discoveries by Fermat on the theory of numbers, de maximis et minimis.
„Diese Ausgabe von Bachet de Meziriac, mit den Anmerkungen von Fermat ist die beste Textausgabe". Cantor, I, p. 396.

245. **Dou, J. P.**, Tractaet van de roeden ende landtmaten door Hollant
ende West-Vrieslant, met meer andere plaetsen. Leyden, J. Paetz
Jz., 1629. 4to.   6.—
La marge infér. un peu courte.

246. **Dühring, E.** und **U.**, Neue Grundmittel und Erfindungen zur Analysis, Algebra, Functionsrechnung und zugehörige Geometrie.
Lpz. 1844. d. veau.   3.50

247. **Dürer, A.**, Etliche underricht zu befestigung der Statt, Schloz
und flecken. Nürenberg, 1527. 25 ll.

— **Id.**, Underweysung der Messung mit dem Zirckel und richt-
scheyt, in Linien Ebnen und gantzen Corporen. Nürnberg, H.
Formschneyder, 1538. 94 ll.
Contains a.o. the „alphabet" by Dürer.

— **Id.**, Hierinn sind begriffen vier bücher von menschlicher Proportion. Nürnberg, H. Formschneider, 1528. 132 ll. W. 94 woodcuts.

— **Perspectiva**, Eyn nützlich büchlin und underweisung der kunst
des Messens, mit dem Zirckel, Richtscheidt oder Linial, (schlechter
vnnd begreifflicher, dann Dürers Bücher"). Franckfort, C. J.
zum Bart, 1546. 45 ll. W. numerous woodcuts.

— **Lencker, H.**, Perspectiva hierinnen auffs kürtzte beschrieben,
mit exempeln eröffnet etc. Nürnberg, D. Gerlatz, 1571. 37 ll. W.
woodcuts.
First edition with the rare leaves AA and AAij which are often lacking.
Tog. 5 vols. in l. sm. fol. limp vellum. 900.—
Very good copy. From the library of the Duke of Brunsvic-Luneburg.

248. **Eddington, A. S.**, The mathematical theory of relativity. Cambridge, 1923. toile.   8.—

249. **Erdmann, B.**, Die Axiome der Geometrie. Lpz. 1877.   4.50

250. **Ernalsteen, J. A. U.**, Joannes Stadius Leonnouthesius, 1527–
1579. Brecht, 1927. Av. portrait, 1 pl. et ill.   2.—
Biographie et bibliographie du célèbre mathématicien néerlandais, pro-
fesseur à l'université de Louvain.

251. **Euclides**, Elementorum ll. XV. Acc. l. XVI. De solidorum regula-
rium comparatione. Auct. C. Clavio S. J. Francof. 1607. 2 tom.
1 vol. Av. de nombr. figg. veau.   5.—
Un coin du titre restauré.

252. — De gli elementi ll. XV. Con gli scholii antichi. Trad. prima
in lingva latina da F. Commandino et hora nella nostra vulgare.
Urbino, D. Frisolini, 1575. Av. de nombr. figg. fol. vél. (Rel. endomm.).   15.—

253. — Zes eerste boekken, van de beginselen der wiskonsten. Vert.
d. J. Wz. Verrooten van Haerlem. En voor hem gedrukt tot Hamburg, Bij Hendrik Werner, 1638. Av. figg. 4to. vél.   18.—
Cette traduction hollandaise, parue à Hambourg, est fort intéressante
comme preuve de l'importance de la langue néerlandaise au 17e siècle.
L'ouvrage contient à la fin (pp. 313–344): Aenhang. Van de duitsche
(lisez: Nederlandsche) spraek.

254. **Euler, L.**, Theorie der Bewegung fester oder starrer Körper.
Hrsg. von J. Ph. Wolfers. Greifswald, 1853. Av. 9 pl. d. veau. 5.—

255. **Euler, T.**, Institutio calculi integralis. Petropoli, 1768–1845. 4 vol.
Av. pl. 4to. veau, dos dor. (t. IV br., n. r.)   24.—
Ex. avec le 4e vol. qui manque souvent.

Prices are in Dutch guilders

256. **E(versdyck), C. F.**, Pagt-tafelen, waar door alle on-ervarene in de rekenkonste, konnen calculeren de pagt, huere, onkost en koop van lande. Mitsg. instructie om te vinden een yders competentie, in alle verhueringen van tienden, molens, visscheryen, etc. Voorders om te berekenen de weerde van alle goude en silvere specien van gelde, etc. Middelburg, S. Clement, 1713. 4to. vél. 7.50

257. **Faa de Bruno, F.**, Traité élémentaire du calcul des erreurs, avec des tables stéréotypées. Paris, 1869. 2.—

258. **Feliciano da Lazesio, F.**, Libro di arithmetica e geometria speculatiua e praticale. Novam. stamp. Vinegia, F. di Alessandro Bindoni, e M. Pasini, 1545. Av. figg. 4to. vél. 25.—

259. **Fergusson, J. J.**, Tafelen van 't geen elck ende een yegelijck van de Seven Vereen. Provincien.... moet contribueren. Midtsg..... om het selve sonder behulp van tafelen te calculeeren. 's Grav., J. Scheltus, 1675. 4to. vél. 7.50

260. **Fiammelli, G. F.**, La riga matematica, dove si tratta del misurare con la vista di lontano senza instrumenti. Roma, 1605. Av. figg. 4to. cart. 7.50

261. **Fries, J. F.**, Die mathematische Naturphilosophie. Heidelberg, 1822. d. rel. 18.—

262. **Friesenborch, H.**, Arithmetica, dat is, ein nye und wolgegründet Rekenboeck. M. vielen Exemplen, so tho deser tydt im Koephandel am gebrucklicksten. Sampt einem geometr.... und astronom. Appendix.... dergelyken vorhen in disser sprake nicht gesehen worden. Embden, 1646. W. figg. on 1 pl. sm. 8vo. vellum. 32.—
Arithmetic book in low-german, very scarce. A corner of the binding damaged.

263. **Gabrilovitsch, L.**, Über mathematisches Denken und den Begriff der aktuellen Form. Berlin, 1914. 1.—
Bibliothek für Philosophie. 8.

264. **Gerhardt, C. J.**, Geschichte der Mathematik in Deutschland. München, 1877. d. rel. 15.—

265. **Graaf, A. de**, Wiskonstige arithmetica, zynde een korte verhandeling van de natuur der cyfer-konst. Op dese order gebragt d. A. van Dam. Amst., G. de Groot, 1696. — **Id.**, Exemplaar-boekje van de arithmetica, zynde een vervolg van de „Wiskonstige Arithmetica", inhoud. alleenlyk die regels, en vraagstukken, die in de negotie voorvallen. Op dese order gebragt d. A. van Dam. Amst., G. de Groot, 1715. — En 1 vol. pet. in-8vo. vél. 12.—

266. — De geheele mathesis of wiskonst, herstelt in zijne natuurlyke gedaante. Amst. 1708. Av. pl. 4to. vél. 9.—

267. **Grandi, G.**, De infinitis infinitorum, et infinite parvorum ordinibus disquisitio geometrica. Pisa, 1710. Av. portrait et figg. — **Id.**, Prostasis ad exceptiones Cl. Varignonii libro de infinitibus infinitorum ordinibus oppositas. Pisa, 1713. Av. figg. — En 1 vol. 4to. vél. 15.—

268. **'s Gravesande, G. J.**, Physices elementa mathematica, experimentis confirmata s. introductio ad philosophiam Newtonianam. Ed. 3a duplo auctior. Leidae, J. A. Langerak et J. et H. Verbeek, 1742. 2 vol. Av. 127 pl. 4to. d. veau. 20.—

269. **Guyot,** Nouvelles récréations physiques et mathematiques, conten. toutes celles qui ont été decouvertes.... sur l'aiman, les
nombres, l'optique, la chymie, etc. Paris, 1769. 4 vol. Av. de nombr. pl. color. d. veau.   15.—

270. **Haan, D. Bierens de,** Bouwstoffen voor de geschiedenis der wis-
en natuurkundige wetenschappen in de Nederlanden. Amst. 1878.
T. I. Av. pl.   20.—
   Contributions à l'histoire des sciences mathématiques et naturelles dans
   les Pays-Bas, surtout dans les 15e aux 18e siècles. Ces études sont de grande
   importance aussi par la description bibliographique très détaillée de plusieurs ouvrages anciens, qui pour la plupart, sont d'une grande rareté.
   Publiées d'abord dans les ,,Verslagen en mededeelingen de Kon. Academie van Wetenschappen", l'auteur en a fait tirer à part 50 exx. qu'il a
   fourni de titres et de tables.
   Pas dans le commerce. Fort rare.

271. **Hage, J.,** Koerstafels. 's-Grav. 1938. cloth.   10.—

272. **Hamilton, W. R.,** Elemente der Quaternionen. Hrsg. von W. E.
Hamilton. Deutsch von P. Glan. Lpz. 1882, 84. 2 tom. 3 vol. d.
veau et br.   12.—

273. **Hellingwerf, P.,** Wiskonstige oeffening behels. eene verhandeling
over veele voorname zaken van de mathesis: De weegkonst, musica, zonnewijzers, vloeystoffen, weegkunde, en mathemat. men-
gelstoffen. Amst. 1718. Av. pl. 4to. vél.   7.50

274. **Herbart, J. F.,** Ueber die Möglichkeit und Notwendigkeit, Mathematik auf Psychologie anzuwenden. Königsbergen, 1822. cart. 5.—

275. **Huygens, Chr.,** Oeuvres complètes. Publ. par la Société Hollandaise des Sciences. La Haye, 1888–1937. T. 1–19. 20 vol. Av. portr.,
pl., facs. et figg. 4to. (355.—)   275.—
   1–10. Correspondance, 1638–95. — 11–13. Travaux mathématiques,
   1645–51; travaux mathématiques pures, 1652–56; dioptrique, 1653, 1666,
   1685–92. — 14. Calcul des probabilités, travaux de mathématiques pures,
   1655–66. — 15. Observations astronomiques. Système de Saturne. Travaux
   astronomiques, 1658–66. — 16. Percussion. Question de l'existence et de la
   perceptibilité du mouvement absolu; etc. — 17. L'horloge à pendule de 1651
   à 1666. Travaux divers de physique, de mécanique et de technique de
   1650 à 1666; etc. — 18. L'horlogerie à pendule ou à balancier de 1666 à
   1695. Anecdota. — 19. Mécanique théorique et physique de 1666 à 1695.

276. — Opera. Lugd. Bat. et Amst. 1724–28. 7 vols. in 3. W. portrait
and 114 pl. 4to. calf and hfcalf.   125.—
   The first 2 vols.: Opera varia. L. B., Janssonius van der Aa, 1724, contain: I. Horologium; II. Geometrica; III. Astronomia; IV. Opera miscellanea.
   The 3d and 4th vol. are entitled: Opera reliqua. Amst., J. Waesbergen,
   1728. Vol. I contains: Tractatus de lumine: De causa gravitatis etc. The
   2d vol. of the Opera reliqua is entitled: Opuscula posthuma. Amst., J.
   Waesbergen, 1728. 2 vols., contain.: I. Dioptrica; II. De coronis et par-
   heliis. De motu et vi centrifuga. Descriptio automati planetarii.
   C o l l e c t i o n   w h i c h   i s   r a r e l y   t o   b e   f o u n d   c o m p l e t e.

277. — Opera varia. L. B. 1724. Vol. I–IV. Av. de nombr. pl. 4to.15.—
   I. Horologium. — II. Geometria. — III. Astronomia. — IV. Opera
   miscellanea.
   Le portrait et un titre manquent.

278. — Horologium oscillatorium sive de motu pendulorum ad horologia aptato demonstrationes geometricae. Paris. 1673. W. figg.
fol. old calf, back gilt. *Fine copy.*   275.—
   O r i g i n a l   e d i t i o n. Chr. Huygens was the inventor of the pendu-

lum-clocks. After having published, in 1658, a smaller treatise, in which he gives a short discussion about his invention, he deals with it in detail for the first time in his „Horologium oscillatorium".
Murhard, IV, p. 186: „Ein unsterbliches Werk".

279. **Jamblichus, Chalcidensis,** De vita Pythagorae, et Protrepticae orationes ad philosophiam ll. II. Gr. et Lat. primum ed. cum notis. In Bibliopolio Commeliniano, 1598. 4to. cart.   45.—

280. **Kästner, A. G.,** Geschichte der Mathematik. Göttingen, 1796–1800. 4 vol. d. veau.   40.—

281. **Keckermann, B.,** Systema compendiosum totius mathematices, hoc est geometriae, opticae, astronomiae et geographiae. Acc. commentatio nautica ab eodem autore. Item methodus facilis arithmeticae practicae per Gemmam Frisium. Oxonii, 1661. Av. 6 pl. pet. in-8vo. vél.   15.—
Dans la même reliure: **Id.,** Systema astronomiae compend. Hanon. 1617. Un peu taché d'eau.

282. **Klein, F.,** Gesammelte mathematische Abhandlungen. Hrsg. von R. Fricke und A. Ostrowski. Berlin, 1921, 22. T. I, II. 2 vol. Av. portrait.   16.—
I. Liniengeometrie. Grundlegung der Geometrie. Zum Erlanger Programm. — II. Anschauliche Geometrie. Substitutionsgruppen und Gleichungstheorie. Zur mathemat. Physik.

283. **Klügel, G. S.,** Mathematisches Wörterbuch, oder Erklärung der Begriffe, Lehrsätze, Aufgaben und Methoden der Mathematik. Lpz. 1803–36. 5 vol. Av. les 2 vol. suppl. Ens. 7 vol.   20.—

284. **Lambert, J. H.,** Anlage zur Architectonic, oder Theorie des Einfachen und des Ersten in der philosoph. und mathemat. Erkenntniss. Riga, 1771. 2 vol. cart.   36.—
Edition originale.

285. — Kurzgefasste Regeln zu perspectivischen Zeichnungen. Augsburg, 1772. Av. 2 pl. cart.   6.—

286. **Landré, C. L.,** Mathemat.-technische Kapitel z. Lebensversicherung. 2e verm. Aufl. Jena, 1901. toile. (6.—)   3.—

287. **Laurent, H.,** Traité d'analyse. Paris, 1885–91. 7 vol. d. veau. 20.—

288. **Leibnitz, G. G.,** Opera omnia. Nunc primum collecta, etc. studio L. Dutens. Genevae, De Tournes, 1768. 6 tom. 7 vol. Av. portrait de L. p. P. Savart et 41 pl., dont 12 de fossiles. 4to. veau fauve.   120.—
Edition recherchée.
Bel ex., frappé aux dos des armes de La Rochefoucauld.

289. — et **J. Bernoulli,** Commercium philosophicum et mathematicum. Laus. 1745. 2 vol. 4to. cart.   18.—

290. **Leslie, J.,** The philosophy of arithmetic. Edinburgh, 1817. d. rel.   6.—

291. **Lilienthal, R. von,** Vorlesungen über Differentialgeometrie. Lpz. 1908, 13. T. I, II. 2 vol. toile et br.   8.—
Tout ce qui a paru.

292. **Lorentz, H. A.,** Lehrbuch der Differential- und Integralrechnung Nebst Einführ. in andere Teile der Mathematik. Übers. von G. C. Schmidt. 4e Aufl. Lpz. 1922. Av. 122 figg. toile. (11.40)   3.50

293. **Marie, M.,** Histoire des sciences mathématiques et physiques. Paris, 1883–88. 12 vol.   24.—

294. **(Marinoni, J. J.)**, De re ichnographia, cujus hodierna praxis exponitur. Vienna, 1751. Av. front., pl. et ill. gr. in-4to. veau. 7.50

295. **Mayr, A.**, Untersuch. über die wissenschaftl. Methode, mit besond. Anwendung a. d. Mathematik. Würzburg, 1845. cart. 3.—

296. **Meyer, G. F.**, Doctrina triangulorum, s. trigonometria, die Lehr von Messung der Trianglen, sambt dem Gebrauch der tabularum, sinuum, tangentium et secantium. (Basel), 1678. Av. 38 grav. à l'eauforte. 12mo-obl. vél. 20.—

297. **Mithobius, Burchardus**, Stereometria, docens certas dimensiones corporum solidorum. Francof., Chr. Egenolphus, 1544. Av. figg., grav. s. bois. pet. in-8vo. cart. 9.—

298. **Molenbroeck, P.**, Theorie der Quaternionen. Leiden, 1891. 2.50

299. **Moor(e), J.**, Mathematical compendium or useful practices in arithmetick, geometry, and astronomy: geography and navigation. 3d ed. London, 1695. Av. 3 pl. 12mo. veau, dos doré. 24.—
Le titre trop court, sans nuire au texte.

300. **Newton, L.**, La méthode des fluxions et des suites infinies. Paris, 1740. 4to. veau. 6.—

301. — Opuscula mathematica, philosophica et philologica. Coll. part. Lat. vertit ac rec. Joh. Castillionens. Lausanne, 1744. 3 vol. 4to. veau. 45.—
Première édition.

302. — Mathematische Prinzipien der Naturlehre. Mit Bemerk. und Erläut. hrsg. von J. Ph. Wolfers. Berlin, 1872. d. veau. *Very scarce.* 30.—

303. **Nierop, D. R. van**, Mathematische calculatie, dat is, wiskonstige rekening: leerende het vinden van verscheyden hemelloopsche voorstellen.... als mede 't beschrijven der zonnewijzers.... bygevoeght de wiskonstige musyka. Amst. 1659. 2 tom. 1 vol. Av. portr. sur les titres et de nombr. figg. et tabl. dans le texte. vél. 20.—

304. **Nieuwentijt, B.**, Gronden van zekerheid of de regte betoogwijse der wiskundigen so in het denkbeeldige als in het zakelijke, ter wederlegging van Spinosaas denkbeeldig samenstel. Amst. 1720. 4to. veau. 20.—

305. — Même ouvrage. Amst. 1739. 4to. vél. 24.—

306. **Nuis, H. J.**, 't Gebruik van het rectangulum cath. geometr.-astronomicum ofte regthoekig meet- en sterkundig plat. Zwolle, G. Tideman, 1686. Av. figg. 4to. vél. 6.—

307. **Ozanam**, Récréations mathématiques et physiques, qui contiennent plusieurs problêmes d'arithmétique, de géometrie, de musique, d'optique, de cosmographie.... Avec un traité des horloges élémentaires. Nouv. éd. revue. Paris, 1735–36. 3 vol. Av. de nombr. pl. et figg. veau. 10.—

308. **Parent, M.**, Essais et recherches de mathématique et de physique. Nouv. éd. augm. d'un 3e vol. Paris, 1713. 3 vol. Av. 31 pl. pet. in-8vo. vél. 18.—

309. **Philolaos** des Pythagoreers Lehren nebst den Bruchstücken seines Werkes. Hrsg. von A. Boeckh. Berlin, 1819. 3.50

310. **Polak, J. F.**, Mathesis forensis, w. i. die Rechenkunst, Geometrie, Baukunst, Mechanik, Hydrostatik und Chronologie und die Anwendung in der Rechtsgelehrsamheit abgehandelt wird. 4e Aufl. Lpz. 1770. Av. portr. et pl. 4to. d. veau. 5.—

311. **Quetelet, A.**, Sciences mathématiques et physiques chez les Belges, au commencement du XIXe siècle. Brux. 1866. cart. 7.50

312. — Histoire des sciences mathématiques et physiques chez les Belges. Nouv. éd. Brux. 1871. 15.—
Epuisé et rare.

313. **Rees, M. van**, Analytische voortzetting van analytische functies en van reeksen van analytische functies. Amst. 1928. 1.—

314. **(Reyneau, R. P.)**, La science du calcul des grandeurs en général ou les élémens des mathématiques. Paris, 1714. 4to. d. veau. 3.—

315. **Ritter, W.**, Anwendungen der graphischen Statik. Nach C. Culmann bearb. Zürich, 1888–1906. 4 vol. Av. pl. et figg. toile. 15.—
I. Die im Inneren eines Balkens wirkenden Kräfte. — II. Das Fachwerk. — III. Der kontinuierliche Balken. — IV. Der Bogen.

316. **Sarganeck, G.**, Die Geometrie in Tabellen. Halle, 1739. 4to-obl. d. veau. 6.—

317. **Scheffers, G.**, Einführung in die Theorie der Kurven in der Ebene und im Raume. Lpz. 1910. 4.50

318. — Lehrbuch der Mathematik. 4e Aufl. Berlin, 1919. Av. 438 figg. (9.—) 6.—

319. **Schooten, Fr. van**, Tabulae sinuum, tangentium et secantium. Gecorr. en bijgev. d'ontbindinge der sphaer. triangulen d. I. I. Stampioen. Rott., Wed. M. Bastiaensz, 1632. Av. titre gravé et figg. 12mo. vél. 6.—

320. **Schott, G.**, Organum mathematicum ll. IX explicatum. Herbip. 1668. W. 48 pl. 4to. calf. 20.—
I. Arithmetica. — II. Geometrica. — III. Fortificatio. — IV. Chronologia. — V. Horographia. — VI. Astronomia. — VII. Astrologia. — VIII. Steganographia. — IX. Musica.

321. **Schwarz, H.**, Versuch einer Philosophie der Mathematik. Halle, 1843. d. veau. 5.—

322. **Simson, R.**, Opera quaedam reliqua, scilicet, Apollonii Pergaei de sectione determinata, porismatum liber, de logarithmis, de limitibus quantitatum et rationum etc. Glasgow, 1776. W. portrait. 4to. calf. (Binding broken). 12.—

323. **Slusius, R. F.**, Mesolabum seu duae mediae proportionales inter extremas datas per circulum et per infinitas hyperbolas, vel ellipses exhibitur. Leodii, 1668. W. figg. 4to. boards, uncut. 40.—
„Slusius avait trouvé sa méthode des Tagentes avant d'avoir vu celle de Newton qui loue beaucoup celle de Slusius".

324. **Speyert van der Eyk, S.**, Beginselen der differentiaal- en integraalrekening. Leyden, 1803. Av. 1 pl. 4to. 1.50

325. **Spinoza, B. de**, Stelkonstige reeckening van den regenboog en reeckening van kansen. 's-Grav. 1883. *T. à p.* 1.50

326. **Stevin, H.**, Wisconstich filosofisch bedrijf, begrepen in veertien

boeken. M. aanh. Leyden, Ph. de Croy, 1667. 15 parts in 1 vol. W. figg. 4to. vellum. W. „Plaetboec", contain. 29 pl. fol.   **95.—**

The works by H. Stevin, son of the famous Simon Stevin, Copies with the 15th part are scarce.

Slightly waterstained. The leaf, contain. the explanations of the plates is in ms.

327. **Stevin, S.,** Oeuvres mathématiques où sont insérées les mémoires mathématiques esquelles s'est exercé le prince Maurice de Nassau. Le tout reveu par A. Girard. Leyde, B. et A. Elsevier, 1634. 2 tom. 1 vol. fol.   **30.—**

Willems, no. 413.

Taches d'humidité.

328. — L'arithmetique. Conten. les computations des nombres arithmetiques ou vulgaires, aussi l'algebre avec les equations de cinc quantitez. Ensemble les quatres premiers livres d'algebre de Diophante d'Alexandrie.... La pratique d'arithmetique, conten. e.a. les tables d'interest, la disme, etc. Leyde, Chr. Plantin, 1585. 2 vols. in 1. W. figg. red mor., gilt edges. (Mod. binding).   **150.—**

First edition. Bibl. belg. S. 129. Smith, Rara Arithmetica p. 385: This work consists of three distinct parts: 1. „L'Arithmetique".... treating of powers and roots; particularly of surds, and of operations on numerical and algebraic expressions and the solution of equations: 2. „Les quatre premiers Livres d'Algebre de Diophante d'Alexandrie", transl. by Stevin, apparently from Xylander's text; 3. „La Pratique d'Arithmetique", an attempt at a practical textbook. — The „Pratique" contains „La Regle d'Interest avec ses tables", „La Disme" etc. The interest centers in „La Disme", in which decimal fractions are for the first time treated in any elaborate way.

The work is dedicated to Jean Cornets de Groot, the father of Hugo Grotius. The dedication contains many interesting particulars concerning the person of Cornets de Groot, who is mentioned there for the first time as a composer. (See Bibl. Belg.).

329. — Materiae politicae. Burgherlicke stoffen. Vervanghende Ghedachtenissen der oeffeninghen des Doorl. Prince Maurits van Orangie, etc..... En uyt sijn naegelate hantschriften by een gestelt door sijn soon Hendrick Stevin. Leyden, Voor A. Roseboom, Schout tot Alphen, (1649). W. 3 pl. and figg. in the text. 4to. vellum.

180.—

This work, which is e x t r e m e l y r a r e, contains a.o.: Onderscheyt van de oirdeningh der steden. M. Byvoug de Huysoirdening (Deals with a new system of founding towns and buildings in squares. The plans resemble those of the first towns of the United States of America, f.i. Philadelphia. In the „Huysoirdening" all details of the houses are dealt with. — Onderscheyt van het burgherlick leven, etc. — Onderscheyt vande ghemeene reghel, op ghesanterie en ghesantsverhael (on the functions of ambassadors, etc.). — Onderscheyt vande crijchspiegheling (Organisation of the army, tactics, etc.). — Verrechting van domeyne, w.o. de vorstelicke bouckhouding op de Itaeliaensche wijse (contains a.o. 213 pp. on bookkeeping).

Contains also 4 ll., not mentioned in the index, contain. a treaty by H. Stevin on the „perpetuum mobile".

330. **Stone,** Analise des infiniments petits compren. le calcul intégral. Trad. p. Rondet. Paris, 1735. Av. pl. 4to. veau.   **6.—**

331. **Struyck, N.,** Oeuvres, qui se rapportent au calcul des chances, à la statistique générale, à la statistique des décès et aux rentes viagères. Trad. du holland. p. J. A. Vollgraff. Amst. 1912. Av. portr. toile. (12.—)   **4.—**

Tirage limité.

Prices are in Dutch guilders

332. **Struyck, N.**, Uytreekeningh der kansen in het speelen, door de arithmetica en algebra, beneevens een verhandeling van looterijen en interest. Amst. 1716. 4to. 7.50

333. **Swinden, J. H. van**, Huygens als uitvinder der slingeruurwerken. Amst. 1817. Av. 5 pl. 4to. *Extr.* 1.50

334. **Tables** de sinus, tangentes, secantes, et logarithmes de sinus, tangentes, et nombres jusques à 10000. Av. methode de resoudre par le moyen d'icelles tous triangles rectilignes et spheriques, etc. Corr. p. A. Vlacq. Leyde, Ph. de Croy, 1651. sm. 8vo. vellum. 15.—

335. **Treutlein, P.**, Geschichte unserer Zahlzeichen und Entwickelung der Ansichten über dieselbe. Karlsruhe, 1875. Av. 1 pl. 1.—
Valentin Mennher de Kempten. — See nr. 681.

336. **Violeine, P. A.**, Nouvelles tables pour les calculs d'intérêts composés d'annuités et d'amortissement. 8e éd. Paris, 1903. 4to. 5.—

337. **Wagner, J. J.**, Mathematische Philosophie. Erlangen, 1811. 6.—

338. **Wallace, W.**, A new book of interest, containing aliquot tables, truly proportioned to any given rate, for the use of the merchant, banker, etc. Wherein is demonstrated, that the tables, in all the common interest books, constantly make the interest less than the true amount, etc. London, 1794. 4to. hfcalf. *Scarce.* 15.—

339. **Westen, W. van**, Mathematische vermaecklyckheden.... arithmetica, geometria, astronomia, geographia, cosmographia, musica, physica, etc. U. h. Fr. Verm. m. annotat. 4e dr. Arnhem, J. van Biesen, 1662. 3 tom. 1 vol. Av. figg. pet.-in-8vo. vél. 12.—
Un peu court de marges.

340. **Wils, P.**, Wis-konstige wercken, bestaende in eenighe meet-konstighe ende hemel-klootsche aenteyckeningen. Amst., Th. Fonteyn, 1654. Av. figg. 4to. *Ex. grand de marges.* 24.—
Voir Bierens de Haan, Bouwstoffen. Très rare. Quelques légères taches.

341. **Witt, Johan de**, Waerdye van lyfrenten naer proportie van losrenten. 's-Grav., J. Scheltus, 1671. fol. boards. 180.—
Original edition, extremely rare, of this treatise written by the Grand Pensionary Johan de Witt in order to propagate the system of life-annuities.
Added: [**Lijfrenten-brief** van 1666. Uitgeg. door de Staten van Hollandt ende West-Vrieslandt. Gedrukt stuk op perkament met de namen en bedragen in hs. Ten bate van Emmerentia Heydanus, oudt 3 jaar, dochter van Prof. A. Heydanus en Corn. Schilperoort. Ondert. door Johan de Wit, Nic. Tulp e.a. c. 40 × 50 cm. Zonder zegel].
See for a similar piece, dated 1665: ,,Een interessant document'' in: ,,Bouwstoffen voor de geschiedenis van de levensverzekeringen en lijfrenten in Nederland'', 1897, p. 237 etc.

342. **Wolf, Chr.**, Elementa matheseos universae. Ed. nova. Genevae, 1740–47. 5 vol. Av. portrait et 163 pl. 4to. vél. 40.—
Traite des mathématiques pures et appliquées, de l'architecture militaire et civile, de l'astronomie, du télescope, etc.

343. **Zubler, L.**, Nova geometria pyrobolia. Neuwe geometr. Büchsenmeistery, wie man jedes Geschütz richten (und) desselben höche und weite mässen soll. Zürich, 1614. 3 tom. Av. titre gravé et 23 grav. en taille-douce. — **Id.**, Novum instrum., geometr., alle Weite, Breite, etc. abzumessen. Basel, 1614. Av. titre gravé et 13 grav. en taille-

douce. — **Id.**, Fabrica et usus instrum. chorograph. D. i. planimetr. Beschreib. wie man alle Stätt, Gärten, etc. auffreissen und verjüngen soll. Basel, 1614. Av. titre gravé et 13 grav. en taille-douce. — En 1 vol. 4to. vél. (Qq piqûres dans les plats.)   30.—

PHYSICS

(See also the division Mathematics)

344. **d'Alembert, J.**, Traité de dynamique. Paris, 1743. 4to. veau. 15.—
    Première édition.

345. — **Condorcet** et **Bossut,** Nouv. expériences sur la résistance des fluides. Paris, 1777. d. veau.   12.—

346. **Borrius, H.**, De motu gravium et levium. Florentiae, G. Marescottus, 1576. Av. portrait de l'auteur, grav. s. bois, au v° du titre. 4to. vél. souple.   12.—
    Les dernières pp. légèr. tachées d'eau.

347. **Bosscha, J.**, Leerboek der natuurkunde. Leiden, 1895–1907. 5 tom. 6 vol. d. rel. (1 vol. toile).   7.50
    Tout ce qui a paru. Tous les vol. en dernières éditions. En partie épuisé.

348. — Verspreide geschriften. Leiden, 1901. 3 vol. Av. portrait et pl. (12.—)   7.50

349. **Boyle, R.**, Experimenta et considerationes de coloribus. 1667. pet. in-8vo. vél.   14.—
    „In Experiment and Consideration touching colours (1663) he described for the first time the iridescence of metallic films and soap-bubbels." (Dict. Nat. Biogr., II, p. 1028.)

350. — Même ouvrage. Gen. 1680. 4to. cart.   7.50

351. — Tractatus. In quibus continent. I. Suspiciones de latentibus quibusdam qualitatibus aeris; c. app. de magnetibus coelet. II. Animadvers. in D. Hobbesii Problemata de vacuo. III. De causa attractionis per suctionem. — **Id.**, Experim. nova circa conservationem corporum in vacuo Boyliano. — Ex Angl. Londini, 1676. 4 parties en 1 vol. 12mo. veau.   7.50

252. **Brugman, A.**, De materia magnetica,ejusque actione in ferrum et magnetem. Leov., H. AE. de Chalmot, 1765. Av. 6 pl. de figg. 4to.   10.—

353. **Cabeus, N.**, Philosophia magnetica in qua magnetis natura explicatur, et omnium quae hoc lapide cernuntur causae afferuntur. Ferrarie, 1629. W. engraved titlepage, numer. figg., engraved on wood and some on copper in the text. fol. boards.   45.—
    Chap. X of the 3d book is entitled: Si causam deviationis à vero meridiano cognovissent, per hanc potuissent Americam certo signo praecognoscere, et praedicere illius orbis inventores.
    On p. 220 a small worldmap, which indicates a.o. Terra Australis incognita.

354. **Cametti, O.**, Mechanica fluidorum s. de aequilibrio et motu corporum fluidorum tractatus. Florent. 1777. Av. 18 pl. se dépliant de figg. 4to. veau.   9.—

355. **Collection, Important,** of physical, chemical, mathematical and technical theses, in the dutch language, written under supervision of the university professors Van Bemmelen, Bierens de Haan,

Prices are in Dutch guilders

Buys Ballot, Cohen, Holleman, Jaeger, Kamerlingh Onnes, Keesom, Korteweg, Kuenen, Lorentz, van der Waals, a.o. and maintained at the universities of Leiden, Utrecht, Groningen, Amsterdam and Delft between the years 1850 and 1926. Together 320 pieces, of which 103 of the 19th, 217 of the 20th century. 300.—

356. **Desaguliers, J. T.**, Lectures of experimental philosophy, where in the principles of mechanicks, hydrostaticks and opticks. London, 1719. Av. 10 pl. 4to. veau. 15.—
L'ouvrage traite de nombr. instruments et expériments, de la théorie de la lumière et des couleurs de Newton, de la machine de Rowley représent. le système solaire, etc. L'ouvrage publié d'abord en nom de D. par P. Dawson sous le titre ,,System of experimental philosophy" fut désavoué par lui.

357. **Droste, J.**, Het zwaartekrachtsveld van een of meer lichamen volgens de theorie van Einstein. Leiden, 1916. 2.—

358. **Eder, J. M.**, und **E. Valenta**, Atlas typischer Spektren. Hrsg v. d. Kais. Akademie der Wissensch. Wien, 1911. Av. 53 pl. gr. in-4to. En portef. toile. 20.—

359. **Fokker, A. D.**, Over Brown'sche bewegingen in het stralingsveld, en waarschijnlijkheids-beschouwingen in de stralingstheorie. Haarlem, 1913. 3.—

360. — Over het magnetisme. 's-Grav. 1939. W. figg. *Reprint*. 1.—

361. **Gordon, J. E. H.**, Physical treatise on electricity and magnetism. London, 1880. 2 vol. Av. pl. toile. 6.—
Edition originale de cet excellent ouvrage.

362. **Guericke, O. de**, Experimenta nova (ut vocantur) Magdeburgica de vacuo spatio. Primum à G. Schotto, S. J., nunc verò ab ipso auctore ed., var. exper. aucta. Acc. De aeris pondere circa terram, de systemate mundi planetario, etc. Amst., J. Janssonius à Waesberge, 1672. W. engraved titlepage, portrait of the author and 20 pl. and ill. fol. Hfcalf. 125.—
One of the greatest books on physics of all time, remarkable for its fine illustrations. Amongst other discoveries Guericke (1602–1686) created a vacuum, a desideratum in science from before Aristotle. ,,This remarkable work ranks next to Gilbert's in the number and importance of the electrical discoveries described". (Weaver, Catal., 1909).

363. **Haas-Lorentz, G. L. de**, De beide hoofdwetten der thermodynamica en hare voornaamste toepassingen. 's-Grav. 1938. cloth. 5.40

364. **Herrero, A. M.**, Physica moderna, experimental, systematica. Madrid, 1738. W. 5 pl. sm. 8vo. limp vellum. 10.—
Huygens, Chr. — See the division Mathematics.

365. **Kircher, A.**, Ars magna lucis et umbrae in X ll. Ed. 2a auctior. Amst., J. Janssonius a Waesberge, 1671. Av. front. (monté), portr., 1 grande pl. et de nombr. figg. mathémat., cosmograph. etc. fol. veau. 20.—

366. **Lambert, J. H.**, Pyrometrie oder vom Maasse des Feuers und der Wärme. Berlin, 1779. Av. 8 pl. 4to. cart. 15.—

367. **(Lasseré, Fr.)**, La dioptrique oculaire, ou la théorique, la positive, et la méchanique de l'oculaire dioptrique en toutes ses especes. Paris, 1671. W. 57 pl. and ill. fol. calf. 25.—
The frontispiece in missing.

368. **Lorentz, H. A.,** Collected papers. (Ed. by P. Zeeman and A. D. Fokker). The Hague, 1934-39. 9 vol. Av. portrait. buckram, tête dor. 120.—

„The edition of the various papers of the world-famed physician collects all treatises which had not already appeared in bookform. With the exception of his dissertation, which is publ. in two languages, the scientific papers are publ. in one of the leading international tongues".

369. — Même ouvrage. 9 vol. br. 100.—

370. — Het verband tusschen de voortplantingssnelheid van het licht en de dichtheid en samenstelling der middelstoffen. Amst. 1878. 4to. cart. (Akad.). 2.—

371. — De wegen der theoret. natuurkunde. Amst. 1905. 1.—
Pas dans le commerce.

372. **Lorenz, R.,** Raumerfüllung und Ionenbeweglichkeit. Lpz. 1922. Av. 1 pl. et 17 figg. 4.50

373. **Loring, F. H.,** Atomic theories. London, 1921. Av. 66 figg. toile. (7.50) 3.50

374. **Lungo, C. del,** Elementi della teoria cinetica dei gas. Bologna, 1920. 2.—

375. **Mairan, De,** Traité physique et histor. de l'aurore boréale. 2e éd. Paris, 1754. Av. 17 pl. 4to. veau. 6.—

376. **Mariotte,** Oeuvres. La Haye, J. Neaulme, 1740. 2 tom. 1 vol. 4to. veau. 7.50
Traité de la percussion ou choc des corps; De la nature de l'air; Des couleurs; Du mouvement des eaux.

377. **Marum, M. van,** Première continuation des expériences, faites par le moyen de la machine électrique Teylerienne. (French and Dutch text). Harlem, 1787. W. 12 pl., 9 of which p r i n t e d i n c o l-o u r s. 4to. Hfcalf. *Scarce.* 45.—

M. van Marum, 1750-1837, famous physicist and, since 1781, director of „Teyler's Genootschap" was especially known for his experiments and inventions in the field of electricity. The present work contains a.o. a beautiful folding plate, representing the electric engine of the Museum Teyler. The 9 superb plates, printed in colours, represent the condition of the different metals after their calcination by electrical unloading. See Kuenen, Het aandeel van Nederland in de ontwikkeling der natuurkunde.
Publ. in: Verhandelingen Teyler's Tweede Genootschap.

378. **Maxwell, J. C.,** Treatise on electricity and magnetism. Oxford, 1873. 2 vol. toile. 6.—

379. **Natuurkundig laboratorium, Het,** der Rijks-Universiteit te Leiden, 1882-1904. Gedenkboek aangeboden aan H. Kamerlingh Onnes, directeur van het laboratorium. Leiden, 1904. Av. portr. et pl. toile. 10.—
Contient des contributions de J. D. van der Waals, P. Zeeman, J. P. Kuenen, J. Bosscha, H. A. Lorentz, e.a.
Pas dans le commerce et fort rare.

381. **Nollet,** L'art des expériences ou avis aux amateurs de la physique sur le choix, la construction et l'usage des instruments. Nouv. éd. Amst., D. J. Changuion, 1770. 3 vol. Av. de nombr. pl. d. veau. 15.—

382. — Leçons de physique expérimentale. Nouv. éd. Paris, 1784. 6 vol. Av. portrait et de nombr. pl. d. veau. 28.—
Contient e.a.: Loix du mouvement. — Gravité — Hydrostatique. — Méchanique. — Nature du feu. — Lumière. — Mouvements des astres. — Electricité. — etc.

Prices are in Dutch guilders

PHYSICS   29

383. **Oersted, H. C.**, The discovery of electromagnetism made in 1820.
Publ. by A. Larsen. Copenh. 1920. 4to.   1.50
   Text in 6 languages.

383a. **Planck, Max.** [Collection of articles on physics publ. at the occasion
of his 80th birthday]. The Hague, 1938. W. portrait, pl. and figg.
                                                                     4.—
   Contributions by G. J. Sizoo, W. J. de Haas, A. D. Fokker, W. H. Keesom,
   M. J. O. Strutt, K. S. Knol, K. W. Taconis, W. G. Burgers a. o.
   Physica. Voli V, nr. 4.

384. **Poincaré, H.**, Electricité et optique. La lumière et les théories élec-
trodynamiques. 2e éd. Paris, 1901. d. veau.   6.—

385. **Stobaeus, J.**, Eclogarum ll. II: quorum prior physicas, poster.
ethicas complectitur. Gr. ed. interpr. G. Cantero. Antv., Chr.
Plantin, 1575. fol. d. vél.   6.—

386. **Théorie, La,** du rayonnement et les quanta. Rapports et discussions de la réunion tenue à Bruxelles, 1911. Publ. par P. Langevin
et M. de Broglie. Paris, 1912. d. rel.   4.50
   Rapports de H. A. Lorentz, M. Knudsen, H. Kamerlingh Onnes, P.
   Langevin, A. Einstein e.a.

387. **Uylenbroek, P. J.**, De fratribus Christiano atque Constantino Hu-
genio artis dioptricae cultoribus. L. B. 1838. Av. figg. gr. in-4to.
                                                                     1.25

388. **Vleuten, A. van,** Geschiedenis van Avogadro's hypothese. Leiden,
1873. toile.   1.25

389. **Voogd, J.**, Leidsche onderzoekingen over den suprageleidenden toestand van metalen, 1927–1930. Amst. 1931. Av. figg.   3.—

390. **Voordrachten** betr. verlichtingskunde gehouden 1937. 's-Grav.
1939. W. 53 figg.   1.80
   D. Vermeulen, Lichttechnische eigenschappen van glas. — **J. A. van
   Heuven,** De psychologie van het zien. — **L. S. Ornstein,** De dag- en avond-
   verlichting van het Haagsche Gemeentemuseum. — etc.

391. **Webster, A. G.**, The dynamics of particles and of rigid, elastic, and
fluid bodies, being lectures on mathematical physics. Lpz. 1912.
Av. ill. et figg. d. rel.   2.50

391a. **Wiersma, E. C.**, Eenige onderzoekingen over paramagnetisme.
's-Grav. 1932. Av. 26 figg.   3.60

392. **Zeeman, P.**, Verhandelingen over magneto-optische verschijnselen.
Leiden, 1921. Av. portrait et figg.   9.—
   Réimpression de tous les traités du Prof. P. Zeeman sur les phénomènes
   magnéto-optiques, écrits en anglais ou en français. Le premier, traitant
   la découverte de Z. de l'influence d'un champ magnétique sur la lumière
   émise par un corps, s'y trouve en 4 langues. Epuisé.

392a. — **Zeeman, Pieter.** 1865–25 Mei–1935. Verhandelingen op 25 Mei
1935 aangeboden aan Prof. Dr. P. Zeeman. 's-Grav. 1935. Av. portrait, 4 pl. et des figg.   3.—
   E. Amaldi und **E. Segré,** Einige spektroskop Eigenschaften hochangeregter Atome. — **G. E. Hale,** The magnetic periodicity of sun-spots. — **N.
   Bohr,** Zeeman effect and theory of atomic constitution. — **G. E. Uhlenbeck**
   and **S. Goudsmit,** Statistical energy distributions for a smaller number of
   particles. — et des contributions de T. Mishima et H. Nagaoka, J. Becquerel, E. Cohen, H. A. Kramers e.a.

   Mart. Nijhoff, The Hague — Cat. No. 632

ASTRONOMY, COSMOGRAPHY

393. **Apelt, E. F.,** Die Reformation der Sternkunde. Jena, 1852. 10.—
Première édition.

394. **Apianus, P.,** Cosmographia. Per Gemmam Frisius apud Lovanienses medicum et mathemat. insignem, iam demum ab omnibus vindicata mendis, ac nonnullis quoque locis aucta. Add. eiusdem argumenti libellis ipsius Gemmae Frisii. Antv., J. Withagius, 1574. W. map, movable figg. and woodcuts. 4to. limp vellum. *Wide margins.* 70.—

395. **Bailly,** Histoire de l'astronomie ancienne, depuis son origine jusqu'à l'établissement de l'école d'Alexandrie. 2e éd. Paris, 1781. — **Id.,** Histoire de l'astronomie moderne depuis la fondation de l'école d'Alexandrie jusqu'à 1730. Nouv. éd. Paris, 1785. 3 vol. — Ens. 4 vol. 4to. veau, dos dor. 30.—
Ouvrage estimé.

396. **Bion, N.,** L'usage des globes céleste et terrestre, et des spheres suivant les différens systemes du monde. Préc. d'un traité de cosmographie. Paris, 1761. Av. 48 cartes et pl. veau. 9.—

397. **Birkeland, K.,** On the cause of magnetic storms and the origin of terrestrial magnetism. Christiania, 1908, 13. 2 vol. Av. figg. gr. in-4to. 15.—
The Norwegian Aurora Polar Expedition 1902–1903. Vol. I.

398. **Blaeu, W. Jz.,** Tweevoudigh onderwijs van de hemelsche en aerdsche globen, het een na de meyning van Ptolemeus, het ander na de natuerlijcke stelling van Copernicus. Amst., J. Blaeu, 1647. Av. figg. 4to. vél. 12.50
Légères piqûres dans les marges intér. de qq. ff.

399. — Même ouvrage. Amst., J. Blaeu, 1666. 2 parts in 1. W. figg.
Bound up with:
— **Graaf, A. de,** De starre-kunst leerende de hoedanigheden der beweginge van alle zichtbare hemelsche lighamen. Amst., P. Goos, 1659. W. 8 folding pl. — **Id.,** Redenering wegens de vinding der lengte van Oost en West. (16 pp.).
In 1 vol. 4to. vellum. *Fine copy.* 60.—

400. — Institutio astronomica de usu globorum et sphaerarum caelestium ac terrestrium. Amst., J. et C. Blaeu, 1640. Av. grav. s. bois. d. vél. 10.—

401. — Même ouvrage. Lat. redd. a M. Hortensio. Amst., J. Wolters, 1690. Av. figg. vél. 8.50
Quelques petites piqûres.

402. — Institution astronomique de l'usage des globes et spheres celestes et terrestres, comprise en deux parties, l'une, suivant.... Ptolomée, qui veut que la terre soit immobile; l'autre, selon.... N. Copernicus, qui tient que la terre est mobile. Amst., J. Blaeu, 1669. Av. plus. figg. 4to. vél. 20.—

403. **Blancanus, J.,** Sphaera mundi, seu cosmographia demonstrativa. In qua mundi fabrica, c. novis Tychonis, Kepleri, Galilei adinven-

tis cont. Mutinae, 1635. 2 vols. in 1. W. pl. and woodcuts in the text. fol. vellum. 32.—

Access. Introductio ad geographiam. Apparatus ad mathematicarum studium. Echometria, i.e. geometrica tractatio de Echo. Nouum instrum. ad horologia describenda.

404. **Brown, B.**, Astronomical-atlases, maps and charts. London, 1932. Av. 19 pl. 4to. toile. (11.—) 5.—

405. **Cassini**, Eléments d'astronomie. — Tables astronom. du soleil, de la lune, des planetes, des étoiles fixes, et des satellites de Jupiter et de Saturne. — Paris, 1740. 2 vols. W. 93 figg. on 26 pl. 4to. calf, gilt back. 20.—

Premières éditions.

406. **Dionysius**, Orbis descriptio. Arati astronomicon. Procli sphaera. C. schol. Ceporini. Colon., J. Gymnicus, 1543. pet. in-8vo. vél. 12.—

Texte grec av. comment. en latin.

407. **Dodonaeus, R.**, De sphaera, s. de astronomiae et geographiae principis cosmographica isagoge.... Nunc vero eiusdem recogn. locupletior facta. Antv., Chr. Plantin, 1584. W. 23 woodcuts. sm. 8vo. contempor. calf binding. 40.—

Second revised and slightly augmented edition of ,,Cosmographica in astronomiam et geographiam isagoge. 1584. This book contains (pp. 105–109) a description of the inhabited earth, extracted from the works of Corn. Valerius and gives on p. 109 an extract of the ,,Medea'' tragedy by L. A. Seneca. D. thought he had found proof of Seneca's knowledge of America's existence in the latter's poems. (See Bibl. Belgica).
Extremely scarce.
In the same binding: **A. Piccolomine** in mechanicas quaestiones Aristotelis etc. Venetijs, T. Curtius, 1565. W. figg.
The leaves 17–24 of this work are missing.

408. **Du Hamel, J. B.**, Astronomia physica, seu de luce, natura, et motibus corporum caelestium. Acc. P. Petiti Observationes aliquot eclipsium solis et lunae. Paris. 1660. 2 parties 1 vol. Av. figg. dans le texte. 4to. veau. 12.—

La 2e partie est déreliée.

409. **Euler, L.**, Theoria motuum planetarum et cometarum. Berol. 1744. Av. front. et 4 pl. 4to. d. veau, n.r. 10.—

Première édition, rare.

410. — Novae tabulae lunares singulari methodo constructae. Petropoli, 1772. 5.—

411. **Fabri S. J., H.**, Dialogi physici, in quibus de motu terrae disputatur, marini aestus nova causa proponitur, necnon aquarum et Mercurii supra libellam elevatio examinatur. Lugd. 1665. Av. figg. 4to. vél. 20.—

412. **Ficinus, Marsilius**, Libellus de sole. Nurmberg, C. Tockler, 1502. 4to. (19 pp.). 35.—

Légèrement taché d'eau.

413. **(Focard, J.)**, Paraphrase de l'astrolabe. Les principes de geometrie. La sphere. L'astrolabe, ou, declaration des choses celestes. Le miroir du monde, ou, exposition des parties de la terre. Lyon, J. de Tournes, 1546. W. many woodcuts. sm. 8vo. gilt calf. 175.—

Original edition. At the end the longitude and latitude of the principal towns and points in Europe, Asia and Africa.
Several ms. annotations.

414. **Galilaei, G.,** Systema cosmicum. Lugd. 1641. W. front., portrait and figg. 4to.   42.50
415. **Gassendus, P.,** Tychonis Brahei vita. Acc. N. Copernici, G. Peurbachii et J. Regiomontani astronomorum celebrium vita. Ed. 2a auctior. Hag. Com., A. Vlacq, 1655. W. 2 portr. and figg. 4to. vellum.   75.—
Library stamp on titlepage.
416. — Institutio astronomica, juxta hypotheseis tam veterum, quam Copernici et Tychonis. Londini, 1675. veau.   10.—
417. — Même ouvrage. Acced. ejusdem varii tractatus astronomici. Amst. 1680. 4to. vél.   15.—
418. **Gemma Frisius,** De astrolabo catholico liber. Antv., J. Grapheus for J. Steelsius, 1556. W. 2 pl. and numer. figg., engraved on wood. sm. 8vo. vellum.   48.—
First edition. See Van Ortroy, Bio-bibliographie de G. F. 1920.
The first cover of the binding slightly damaged, otherwise fine copy.
419. — The same work. Antv., J. Withagius, 1583. W. vign. on titlepage and figg. in the text. 4to. hfmor. (Mod. binding).   25.—
420. **Gruithuisen, F. v. P.,** Die Natur der Kometen, mit Reflexionen auf ihre Bewohnbarkeit und Schicksale; bey Gelegenheit des Kometen von 1811. München, 1811. W. 4 pl. sm. 8vo. cloth.   7.50
421. **Gunter, E.,** Works, contein. the description and use of the sector, cross-staff, a.o. instruments, with a canon of artificiall signes and tangents .... and the use there of in astronomie, navigation, dialling, etc. Enl. with a treatise on fortification. W. the further use of the quadrant fitted for daily practise by S. Foster. 3d ed. amended by H. Bond. London, 1653, 52. W. engraved titlepage, 1 pl. and figg. 4to. ancient calf.   50.—
422. **H(arris), J.,** Astronomical dialogues between a gentleman and a lady, wherein the doctrine of the sphere, use of the globes, and the elements of astronomy and geography are explain'd. W. descriptions of the famous instrument call'd the orrery. 2d ed. London, 1725. W. 5 pl. calf.   18.—
423. **Helfrecht, J. T. B.,** Tycho Brahe, geschildert nach seinem Leben, Meynungen und Schriften. Hof, 1798. Av. 1 pl.   6.—
Sans le portrait.
424. **Hemminga, S. ab,** Astrologiae, ratione et experientia refutatae liber: cont. brevem apodixin de incertitudine astrologica edita contra astrologos Cyper. Leouitium, H. Cardanum et L. Gauricum. Antv., Chr. Plantin, 1583. W. figg. 4to. limp vellum. (Back damaged).   25.—
425. **Herschel, W.,** Ueber den Bau des Himmels. Drey Abhandlungen. A. d. Engl. Nebst einem authent. Auszug aus Kants allgem. Naturgeschichte und Theorie des Himmels. Königsbergen, 1791. cart. *Rare.*   22.50
426. **Hess, W.,** Himmels- und Naturerscheinungen in Einblattdrucken des 15. bis 18. Jahrh. Lpz. 1911. Av. 30 pl., dont qq.-unes en couleurs. 4to.   5.—
Tiré à petit nombre.

Prices are in Dutch guilders

427. **Hevelius, J.**, Selenographia, s., lunae descriptio; atque accurata, tam macularum ejus, quam motuum diversorum.... delineatio. Gedani (Danzig), 1647. W. front., portrait and 89 pl. fol. vellum. *Fine copy.* 225.—
Classical work. ,,Am meisten Werth hat seine Selenographie und er (Hevelius) ist als der Vater der Mondbeschreibung zu bezeichnen''. (Allgem. deutsche Biographie).
Stamp on titlepage.
Bound up with: **Hevelius, J.,** Eclipsis solis observata. Gedani, 1649. — Idem. 1652. — Tog. 6 ll. of text with 3 pl., of which one in the text.

428. **Honterus, J.**, Rudimentorum cosmographicorum ll. III cum tabellis geographicis elegantissimis. De variarum rerum nomenclaturis per classes, l. I. No pl. 1581. Titlepage anciently coloured, 29 ll. of text and 2 blank ll., 14 anciently coloured ll., containing maps and a sphere and 2 blank ll. at the end. sm. 8vo. limp vellum. *Fine copy.* 125.—
At the bottom of the first map ,,Universalis cosmographia'' is engraved: Tiguri MDXLVI and the monogram H.V.E.
,,In this map there is, on the west, a continent named America, and a narrow strip, separated from the latter, with the word Parias''. (Harrisse, Bibl. Amer., p. 432).
Some words in ancient ms. on the titlepage.

429. **Hues, R.**, Tractaet, ofte handelinge van het gebruyck der hemelscher ende aertscher globe. Nu in Nederduytsch.... met diversche nieuwe verklaringen verm..... d. J. I. Pontanum. Amst., J. Hondius, 1622. W. engraving on titlepage and figg. in the text. 4to. vellum. 48.—
Rare edition, not mentioned by Bierens de Haan.

430. **Huygens, Chr.**, Systema Saturnium, s. de causis mirandorum Saturni phaenomenon, et comite ejus planeta novo. Hag. Com., A. Vlacq, 1659. W. 1 folding plate and engravings in the text. 4to. calf. 140.—
First edition of Huygens' discovery of Saturn's ring and its fourth satellite.
Binding slightly damaged, but a tall copy.

431. — Kosmotheoros, s. de terris coelestibus, earumque ornatu, conjecturae. Hag. Com., A. Moetjens, 1698. Av. 5 pl. 4to. boards. 40.—
First edition.

432. — Même ouvrage. Ed. 2a. H. C., A. Moetjens, 1699. Av. 5 pl. 4to. veau.*Bel ex. 25.—

433. — Même ouvrage. Ed. 2a. Francof. 1704. Av. 5 pl. se dépliant. pet. in-8vo. vél. aux armes. 15.—
Ex. interfolié de papier blanc.

434. — The celestial worlds discover'd, or, conjectures concern. the inhabitants, plants and productions of the worlds in the planets. (Fr. the) Lat. London, 1698. W. 4 pl. calf. 20.—
First English edition.

435. — De wereld-beschouwer, of gissingen over de hemelsche aardklooten, en derzelver cieraad. U. h. Lat. d. P. Rabus. Rott., B. Bos, 1699. Av. 6 pl. pet. in-8vo. vél. 15.—

436. **Ideler, L.**, Histor. Untersuch. über die astronom. Beobachtungen der Alten. Berlin, 1806. cart. 12.—

437. **Inscriptiones** Hafnienses Latinae Danicae et Germanicae una cum inscriptionibus Amagriensibus Uraniburgius et Stellaeburgicis nec non duabus epistolis T. Brahe ad Peucerum missa.... coll. et cur. P. J. Resenius. Hafnia, 1668. W. 3 pl. 4to. hfcalf.   35.—
From the famous Cortiniana library.

438. **Jacquinot, D.,** L'usaige de l'astrolabe. Avec un traicté de la sphere. Paris, J. Barbé, 1545. W. numer. diagrams. 4to. hfmor.   225.—
The extremely rare first edition.

439. **Jode, C. de, (Corn. de Iudaeis),** De quadrante geometrico libellus, in quo quidquid ad linearum et superficierum utpote altitudinem et latitudinem, dimensiones facit lucidissime demonstratur. Noribergae, Chr. Lochner, 1594. W. engraving on titlepage, 2 coats of arms, 1 pl. and 32 copperengravings in the text. 4to. limp vellum.
125.—
A very scarce and interesting booklet of this famous cartographer on the use of quadrants and theodolites.
See for Corn. de Jode and his works: L'oeuvre cartographique de G. et C. de J. par F. van Ortroy. Gand, 1914, pp. XXIV, etc.
Stamp on titlepage. Slightly browned.

440. **Keckermann, B.,** Systema geographicum. Hanovia, 1611. — **Id.,** Brevis commentatio nautica. Ib. 1611. — **Id.,** Systema astronomiae compendiosum. Ib. 1611. — **Id.,** De natura et proprietatibus historiae. Ib. 1610. — En 1 vol. vél.   28.—

441. **Keppler, J.,** De stella nova in pede Serpentarii et Trigono Igneo. Acc. I: De stella incognita cygni. II: De Jesu Christi servatoris vero anno natalito, consideratio novissimae sententiae Laur. Suslygae. Pragae, 1606. 2 parts in 1 vol. W. pl. 4to. vellum. 110.—
See for the importance of this work: Kästner, Geschichte der Mathematik. IV, pp. 229–235.

442. — Epitome astronomiae Copernicanae usitata forma quaestionum et responsionum conscripta, inque VII libros digesta, quorum tres hi priores sunt de doctrina sphaerica. Lentiis (Lintz a.d. Donau), 1618. — **Id.,** Liber IV.... physica coelestis. Lentiis, 1622. — **Id.,** Libri V–VII.... doctrina theorica. Francof. 1612. — Tog. 3 parts in 1 vol. W. 1 pl. and figg. in the text. sm. 8vo. vellum. 98.—
The 3 parts are seldom found together.

443. **Lansbergius, Ph.,** Bedenckingen, op den daghelijckschen ende jaerlijckschen loop vanden aerdt-cloot. Mitsg. op de ware af-beeldinghe des sienelijcken hemels. Middelburgh, Z. Boman, 1629. Av. fig. sur le titre et dans le texte et 1 pl. 4to. cart.   16.—
Contient une réfutation du système de Tycho Brahe.

444. — Même ouvrage. Middelburgh, W. Goeree, 1675. Av. 1 grande pl. et figg. 4to.   12.—
C'est l'édition de Middelburgh, Z. Roman, 1666, av. un nouveau titre et préface.
Légèrement taché d'eau.

445. — Verklaringe van het gebruyk des astronom. en geometr. quadrants. Middelburg, 1667. Av. figg. et pl. 4to. d. vél.   10.—

446. — Même ouvrage. Nevens het maken des selven quadrants d. M. van Nispen. Verm. dr. Dordrecht, 1685. 4to. d. vél.   10.—

447. **Laplace, P. S. de,** Traité de mécanique céleste. Paris, 1799–1825. 5 vol. 4to. veau.   50.—
Première édition avec tous les 4 suppl.

Prices are in Dutch guilders

448. **Laplace, P. S. de,** Mechanik des Himmels. Übers. von J. C. Burckhardt. Berlin, 1800, 02. 2 vol. 4to.   12.—

449. — Exposition du système du monde. **3e éd.** Paris, 1808. 2 vol. veau.   6.—

450. — Même ouvrage. 5e éd. Paris, 1824. 4to. d. veau. (Dos endomm.).   6.—

451. — Même ouvrage. 6e éd. Paris, 1835. 4to.   12.—
Edition définitive, à laquelle on a ajouté les chapîtres 12, 17 et 18 de livre IV, qui furent supprimés dans la 5e édition.

452. — Darstellung des Weltsystems. Uebers. von J. K. F. Hauff. Frankfurt a. M. 1797. 2 vol. cart.   7.50

453. **Le Bouvier de Fontenelle,** Entretiens sur la pluralité des mondes. Amst. 1686. pet. in-8vo.   12.—

454. **Leçons de géometrie,** pour servir d'introduction à l'étude de la sphère et de la géographie. Paris, 1775. Av. 14 pl. de figg. se dépliant. veau marbré, dos doré.   18.—

455. **Liais, E.,** Traité d'astronomie appliquée à la géographie et à la navigation suivi de la géodosie pratique. Paris, 1867. Av. figg.   5.—

456. **Liesganig, J.,** Dimensio graduum meridiani Viennensis et Hungarici. Vindob. 1770. Av. 10 pl. 4to. cart.   22.50

457. **Maginus, J. A.,** Briefve instruction sur les apparences et admirables effects du miroir concave spherique. Trad. de l'ital. p. J. J. Boyssier. Paris, 1620. Av. figg. dans le texte. 4to. vél. souple.   30.—

458. **Martinus Poblacion, J.,** De usu astrolabi compendium. Paris., J. Barbaeus, 1546. W. woodcuts from the work of Stofler. sm. 8vo. hfcalf. (Mod. binding).   38.—

459. **Maupertuis, M.,** Discours sur la parallaxe de la lune, pour perfectionner la théorie de la lune et celle de la terre. Paris, 1741. veau. (Rel. cassée).   5.—

460. **Mercier, L. S.,** De l'impossibilité du système astronomique de Copernic et de Newton. Paris, 1806. vél.   6.—

461. **Merula, G.,** Nuova selva di varia lettione. Trad. di Lat. Venetia, G. A. Valuassori, 1559. Av. figg. s. bois dans le texte. pet. in-8vo. vél. souple. (Dos endomm.).   10.—
Delle potenze celesti, prima della luna. — Pianeti. — De tori, lupo, leoni, cane. — Torre di Pisa. — Del olio, vino. — Africa, Asia, America. — Come si chiamino i metalli da gli alchimisti. — etc.

462. **Metius, A.,** Institutionum astronomicarum tomi III. ....Acc. de novis autoris instrumentis. Franeker, A. Radaeus, 1608–05. 4 parties en 1 vol. Av. figg. pet. in-8vo. vél. (Rel. légèr. endomm.).   20.—
I. De sphaera. — II. De fabrica planisphaerii et trigonometria astronomica. — III. Historia astronomica. — IV. De novis ab autore invertis instrumentis.

463. — De genuino usu utriusque globi tractatus. Adj. est nova sciatericorum, et artis navigandi ratio novis instrumentis, et inventionibus ill. Franekerae, U. Balck, 1624. W. numer. figg., engraved on wood.

Metius, A., Mensura geographica et usus globi terrestris, artisque navigandi institutio, novis instrumentis et inventionibus adaucta. Franekerae, U. Balck, (1624). W. figg.
Two of the oldest books, publ. in Holland, on nautical instruments and globes.

— **Hues, R.,** Tractatus de globis coelesti et terrestri eorumque usu. Semelque atque iterum à J. Hondio excusus, et nunc elegantibus iconibus et figuris locupletatus ac de novo recognitus.... opera J. I. Pontani. Amst., H. Hondius, 1624. W. 1 pl. and several woodcuts.
Very important work on globes. The additions and corrections by Hondius and Pontanus have made quite a new work of it. Among the woodcuts there is a map of the two Americas and Australia.

                         In 1 vol. 4to. hfcalf. 150.—

464. **Mogge, J.,** Algem. manier tot de practijck-oeffeningh der sonnewysers. Middelburg, W. Goeree, 1675. Av. figg. — **Lansbergen, Ph.,** Beschrijv. der vlacke sonne-wysers. Nieuw. oversien d. J. Mogge. Ibid., id., 1675. Av. figg. — En 1 vol. pet. in-fol. d. vél.   15.—

465. **Moolen, S. van der,** Astronomia of hemelloopkunde, leerende de hoedanigheden der beweginge van alle zigtbaare hemelligten.... de verduysteringe van zon en maan: midsg. het berekenen der tafelen die hier toe noodig zijn.... manier om de tafelen op Mercurius te berekenen.... etc. Amst., J. Loots, 1702. W. front. and 13 pl. 4to. vellum.   12.50

466. **Nierop, D. R. van,** Nederduytsche astronomia, d.i.: onderwijs van den loop des hemels.... Hier by een aen-hangh, dienende tot naeder verklaeringhe over den loop des hemels, als oock eenighe voorbeelden der son-eclipsen. 2e dr. Amst., G. van Goedesbergh, 1658. Av. 2 pl. et figg. dans le texte. 4to. cart.   15.—
Sans le front.

467. — Des aertrycks beweging, en de sonne stilstant, bewijsende dat dit geensins met de Christelijke religie is strijdende.... Met noch verscheyden aenmerck., soo van de vindingh der lenghte van Oost en West, enz. Amst., G. van Goedesbergh, 1661. Av. figg. dans le texte. 4to. cart., n. r.   25.—
Dans la même reliure: **Nierop, D. R. van,** Antwoord op den brief van Jac. Coccejus over de t'samenstellingen des werrelds. Ibid., id., 1661. Av. figg. dans le texte.
La marge supér. des deux ouvrages a un peu souffert d'humidité.

468. **Oefening, Astronomische,** verhandel. de beginselen der sterreloopkunde.... by wyze van vraagen en antwoorden. Amst., erven F. Houttuyn, 1769, 71. 2 tom. 1 vol. Av. 24 pl. color., reprod. 78 figg. astronom. pet. in-8vo. d. veau.   10.—

469. **Pagan, de,** La theorie des planetes, ou tous les orbes celestes sont geometriquement ordonnez; contre le sentiment des astronomes. Paris, 1657. Av. figg. dans le texte. 4to. veau.   10.—

470. **Pannekoek, A.,** Researches on the structure of the universe. 1. The local starsystem deduced from the durchmusterung catalogues. Amst. 1924. W. 6 coloured pl. 4to. sewed.   6.—
Publ. of the astronom. institute of the University of Amsterdam. I.

471. **Petit, P.,** Diss. sur la nature des cometes. Av. discours sur les prognostiques des eclipses etc. Paris, 1665. Av. 2 pl. 4to. veau. 15.—

Prices are in Dutch guilders

472. **Peucer, C.**, Elementa doctrinae de circulis coelestibus, et primo motu. Witebergae, S. Gronenberg, 1587. W. 7 folding tabl., 4 pl. w. movable figg. and numer. astronom. woodcuts. vellum. 35.—
The „Elementa" by Peucer are estimated among the best astronom. works of the 2d half of the 16th century.

473. **Piccolomini, A.**, De la sfera del mondo ll. IV. Novam. emend., e ampl. De le stelle fisse libro. Venetia, G. Varisco, e Comp., 1561. W. astronom. maps on 48 pp. and figg. 4to. vellum. 25.—
First revised edition.

474. — The same work. Accresc., e fino à 6 libri, di 4 che erano ampliata, e quasi per ogni parte rinov. Vinegia, G. Varisco, e P. Paganini, (1564). W. figg. 4to. limp vellum. 35.—
Revised ed., enlarged with 2 chapters.

475. **Pontanus, J. J.**, Centrum Ptolemaei sententiae, a J. J. Pontano e Gr. in Lat. transl. atque expos. — De rebus coelestibus ll. XIIII. Flor., Haer. Philippi Iuntae, 1520. pet. in-8vo. vél. 10.—
Forment les parties 5 et 6 des Opera omnia.

476. **Ptolemaeus**, Liber de analemmate, a F. Commandino instauratus, et comment. ill. Romae, P. Manutius, 1562. Av. figg. dans le texte. 4to. veau. *Bel ex.* 4.50
La sign. G. (= pp. 25—28) manque.

477. **Puchaim, O. F. à**, Diss. physico-mathematica de motu trepidationis terrae. Viennae, 1622. Av. grande pl. gravée. 4to. 20.—

478. **Rivard, M.**, Traité de la sphère. 2e éd. augm. (Av.) traité du calendrier. Paris, 1743, 44. 2 tom. 1 vol. Av. 3 pl. et 1 tabl. veau. 10.—

479. — Même ouvrage. 4e éd. augm. (Av.) traité du calendrier. Paris, 1758. 2 vols. in 1. W. 3 pl. and 1 table. calf. 8.50
Small wormholes.

480. **Sacro Busto, J. de**, De sphaera. C. praef. Ph. Melanthonis. Vittenbergae, J. Crato, 1558. Av. figg., grav. s. bois, dont 1 fig. mobile. pet. in-8vo. dos en peau de truie. 24.—
Ecritures sur le titre et annotat. marginales.
Dans la même reliure: **H. Beyer**, Quaestiones in libellum de sphaera Joannis de Sacro Busto. Francof., P. Brubacchius, 1556. Av. 2 pl., dont 1 un peu défectueuse.

481. **Santarem, de**, Essai sur l'histoire de la cosmographie et de la cartographie pendant le moyen-âge et sur les progrès de la géographie après les grandes découvertes du XVe siècle. Paris, 1849–50. 3 vol. d. veau. *Très rare.* 50.—

482. **Santbech, D.**, Problematum astronomicorum et geometricorum sectiones VII. Basil., H. Petri et P. Perna, 1561. W. numer. woodcuts, a.o. of sun-dials and of canons. fol. limp vellum. 45.—
Contains a.o.: Sectio 1a. Ton phainomenon. — Sectio 2a. De canonibus primi motus. — Sectio 3a. De rationibus gnomonum ac umbrarum ac fundamento sciotericorum instrumentorum. — . . . .Sectio 5a: De ratione librationis, cuius usus est in ducendis aquis. — Sectio 6a. De artificio eiaculandi sphaeras e tormentis. — Sectio 7a. De observat. geographicis.
Chap. CLIIII. Qua ratione navium à littore intervalla tam diei tempore quam in densissimus noctis tenebris per faces ardentes exquisitissimè liceat explorare, etc. — Chap. CLV. Quomodo liceat tam in maritimis, quam terrestribus loeis quisnam ventus quolibet momento spiret, observare.

483. **Scherer S. J., H.**, Critica quadripartita, in qua plura recens inventa et emendata circa geographiae artificium, historiam, technicam, et astrologicam. Monach. 1710. Av. 19 pl. 4to. vél. 30.—

484. **Schotanus à Sterringa, J.,** Physica coelestis, et terrestris. Frane-
     querae, J. Gyzelaar, 1700. pet. in-8vo. veau.   12.—

485. **Schroeter, J. H.,** Selenotopographische Fragmente zur genauern
     Kenntniss der Mondfläche, ihrer erlittenen Veränderungen und
     Atmosphäre. Lilienthal, Göttingen, 1791, 1802. 2 vols. W. 75 pl.
     4to. calf and vellum.   120.—
     Schroeter was the f o u n d e r o f t h e s c i e n t i f i c s e l e n o g r a-
     p h y. The second volume is extremely scarce. (See: Allgem. Deutsche
     Biographie. Bd 32, pp. 570–572).

486. **Stirrup, Th.,** Horometria: or the compleat dialist: wherein the
     whole mystery of the art of dialling is plainty taught three several
     wayes; two of which are performed geometrically by rule and
     compasse onely: and the third instrumentally, by a quadrant
     fitted for that purposes. With the working of such propositions
     of the sphere, as are most usefull in astronomie and navigation,
     both geometrically and instrumentally. London, 1652. W. 1 pl.
     and figg. in the text. 4to. vellum.   75.—

487. **Toaldo, G.,** Della vera influenza degli astri, delle stagioni, e mu-
     tazioni di tempo, saggio meteorologico, applic. agli usi della me-
     dicina, nautica, etc. C. descriz. d'un nuovo pendolo a correzione d.
     Ch. P. Boscovich. Padova, 1770. Av. tabl. et 1 pl. de figg. 4to.
     vél.   15.—

488. **Triegler von Igleraw, J. G.,** Sphaera. Das ist ein kurtzes astronom.
     Tractätlein, in welchem nicht allein von des Himmels Lauff,
     etc., sondern auch, wie man Nativiteten rechnen, und daraus
     urtheilen sol, gehandelt wird. Lpz. 1622. W. the coats of arms
     of Chr. von Dorstad, engraved by A. Bretschneider, at full-page
     size, 1 pl., 1 tabl., numer. astronom. figg. and printer's mark at
     the end. 4to. vellum.   28.—

489. **Tripolita, Th.,** Sphaericorum ll. III, nunquam antehac Graece
     excusi. Iidem Lat. redd. per J. Peram. Paris., A. Wechel, 1558.
     2 parts in 1. W. figg. 4to. calf.   40.—
     In the same binding: **M. Ghetaldus,** Apollonius redivivus. Seu, restituta
     Apollonii Pergaei inclinationum geometria. Venet. 1607. W. figg. — **Id.,**
     Supplementum Apollonii Galli. Seu, exsucitata A. Pergaei tactionum
     geometriae pars reliqua. Venet. 1607. W. figg. — **Id.,** Variorum problema-
     tum collectio. Venet. 1607. W. figg.

490. **Ursus, N. R.,** De astronomicis hypothesibus seu systemate mun-
     dano, tractatus astronomicus et cosmographicus. Item astrono-
     micarum hypothesium a se inventarum, oblatarum, et editorum,
     contra quosdam eas sibi temerario seu potius nefario ausu arrogan-
     tes, vendicatio et defensio. Pragae, apud autorem, 1597. W.
     diagrams. 4to. contemp. limp vellum. *Fine copy.*   120.—
     A very rare and remarkable work, in which the author vehemently
     attacks Tycho de Brahe. Ursus invented a new system of astronomy, very
     little different from that of Tycho Brahe. He communicated it in 1586 to
     the landgrave of Hessen, which gave rise to an angry dispute between him
     and Brahe, who charged him with being a plagiary, while Ursus boasted,
     that he himself was the inventor.

491. **Valk, G.,** 't Werkstellige der sterre-konst mitsg. van de toestand
     des aard-kloots, in 't gebruyck van hemel en aardgloben, benef-
     fens nieuwen planeten-wijzer. Amst. (v. 1700). Av. front. et pl.
     4to. cart.   7.50

                    Prices are in Dutch guilders

492. **Weigel, E.**, Cosmologia nucleum astronomiae et geographiae, ut et usum globorum, tum vulgarium tum novis adornation. etc. tradens. Ed. 2a. Jenae, 1680. Av. 1 pl. se dépliant. 4to. cart.   10.—
493. **Winshemius, S. Th.**, Breve, perspicuum, et facile compendium logisticae astronomicae scriptum. Witteb. 1563. Av. 2 tabl. pet. in-8vo. cart.   12.—
494. **Wohlwill, E.**, Galilei und sein Kampf für die Copernicanische Lehre. Hamburg, 1909. T. I. d. rel.   6.—
Wijk, W. E. van. — S e e n r s. 595–597.

INSTRUMENTS, MACHINES, TECHNICAL WORKS, INDUSTRIES, TRADES.

495. **Aeronautics.** — **5 Pieces.** 1910–14. W. figg.   2.—
　　G. Wellner, Die Flugmaschinen. — **A. Haenig,** Ballon und Flugmotoren. — **C. Eberhardt,** Luftschrauben. — etc.
496. **Ans, Dix,** d'efforts scientifiques et industriels (et coloniaux). (Dir. J. Gérard). 1914–1924. Paris, 1926. 2 forts vol. Av. de nombr. portr. et ill. 4to. toile.   20.—
　　Contient un grand nombre d'articles, écrits par de différ. auteurs, sur la chimie, l'industrie, la technique, l'outillage économique, etc. de la France et des colonies.
497. **Askinson, G. W.**, Perfumes and cosmetics, their preparation and manufacture. London, 1919. Av. figg. toile. (12.50)   5.—
498. **Baratterie, G. B.**, Architettura d'acque. Piacenza, 1656, 63. 2 vol. Av. figg., grav. s. bois. fol. vél.   15.—
499. **Batut, A.**, La photographie appliquée à la production du type, d'une famille, d'un tribu ou d'une race. Paris, 1887. Av. 2 pl.   1.25
500. **Beck, O. W.**, Art principles in portrait photography; composition, treatment of backgrounds and the processes involved in manipulating the plate. N.-York, 1907. W. 138 pl. cloth. (9.50) 3.—
501. **Berthoud, F.**, Essai sur l'horlogerie, rel. à l'usage civil, à l'astronomie et à la navigation. Paris, 1763. 2 vol. Av. 38 pl. 4to. veau marbré.   25.—
502. **Beschrijving, Volledige,** van alle konsten, ambachten, handwerken, fabrieken, trafieken enz., ten deele overgenomen uit.... buitenlandsche werken. Dordrecht, 1788–1820. 24 vols. in 11. W. numerous pl. Hfcalf   150.—
　　Collection of extensive monographs on different trades, such as the paper-maker, the organ-builder, the potter (very important monograph), the engraver, the manufacturer of porcelain, the glas-blower, etc.
　　I. **(Kasteleyn, P. J.),** De indigobereider en blauwverwer. — II. **Demachy,** De sterkwaterstooker, enz., d. P. J. Kasteleyn. — III. **De Milly,** De porceleinfabriek, d. idem. — IV. **Kasteleyn, P. J.,** De leerlooijer en leertouwer, etc. — V. De kaarsenmaker. — VI. **Martinet, J. F.,** Het houtskoolenbranden. — VII. **Macquer,** De zijdeverwer, d. P. J. Kasteleyn. — VIII. **Du Hamel Du Monceau, e.a.,** De zeepsieder, d. idem. — IX. **De la Lande,** De papiermaaker, d. idem. — X. **Du Hamel Du Monceau,** De waschbleeker en waschkaarsenmaker, d. idem. — XI. **Reisig, J. H.,** De suikerraffinadeur. — XII. **Paape, G.,** De plateelbakker. — XIII. **Fokke Sz., A.,** De graveur. — XIV. De honingbijenteelt. — XV. De zijdenteelt. — XVI. **Buys, J.,** De bierbrouwer. — XVII. **Kanter Phz., J. de,** De meekrapteler. — XVIII. **Olivier Schilperoort, T.,** De azijnmaker. — XIX—XXI. **Heurn, J. van,** De

40   INSTRUMENTS, MACHINES, TECHNICAL WORKS, INDUSTRIES, TRADES

orgelmaker. — XXII. **Haas, H. de,** De boekbinder. — XXIII. **(Dalen, van),** De bouwkunstenaar. — XXIV. **Boot, C.,** De verwer.
Very important publication for the knowledge of the arts and professions at the end of the 18th century.
C o m p l e t e   c o l l e c t i o n s   a r e   v e r y   s c a r c e.

503. **Besson, D.,** Teatro de los instrumentos y figuras matemáticas y mecánicas. C. interpr. p. F. Beroaldo. Nuev. impr. Leon de Francia, 1602. Av. titre gravé et 60 ill. en grandeur de la p. fol. d. veau. 45.—
    Traduction espagnole, fort rare, d'un livre célèbre. Les illustr. représentent des instruments mathémat., des machines, comme un banc de tourneur, une machine pour fendre et polir le marbre, un moulin à farine, une charrue, un puits de mine, une roue à puiser, une pompe à feu, etc. enfin des bateaux, un soufflet, des calèches, un violon, etc.

504. **Beyer, J. M.,** Theatrum machinarum molarium oder Schau-Platz der Mühlen-Bau-Kunst. Lpz. 1735. 2 tom. et append. en 1 vol. Av. 43 pl. fol. veau. 45.—
    Forme le t. IX de l'ouvrage de Leupold, Theatrum machinarum generale.

505. **Bibliothèque physico-économique, instructive et amusante,** conten. des mémoires et observat. pratiques sur l'économie rurale, sur les nouvelles découvertes, la description de nouvelles machines et instrumens, etc. Paris, 1782–92. Année 1–11. En 18 vol. Av. pl. pet. in-8vo. veau. 36.—
    Tête de la première série, dirigée par Parmentier et Deyeux. Les pp. 249 à 256 de 1784 manquent.

506. **Bion, N.,** Mathematische Werck-Schule, oder Anweisung wie die mathemat. Instrumenten zu gebrauchen und zu verfertigen. A. d. Franz. Franckfurt, 1712. Av. front. et 28 pl. de figg. 4to. vél. *Bel ex.* 15.—

507. **Bökler, G. A.,** Theatrum machinarum novum, d. i. Neuvermehrter Schauplatz der mechan. Künsten, handelt von allerhand Wasser-Wind.... Mühlen. Nürnberg, 1661. Av. front. et 154 pl. fol. vél. 45.—

508. — Même ouvrage. Nürnberg, 1673. Av. titre gravé et 154 pl. fol. vél. *Ex. très frais.* 45.—
    La dernière pl. représente une pompe à feu, éteignant un incendie.

509. **Borgnis, J. A.,** Traité complet de mécanique appliquée aux arts, conten. l'exposition méthodique des théories et des expériences les plus utiles pour diriger le choix, l'invention, la construction et l'emploi de toutes les espèces de machines. Paris, 1818–23. 10 vol. Av. 249 pl. 4to. veau. *Très bel ex.* 45.—

510. **Bry, H. D. von,** Sonderbahre und biszher verborgen-gewesene Geheime Künste wovon die erste genannt wird: Ort-Forschung, dadurch einer dem andere durch die freye Lufft über Wasser und Land von einem sichtbaren Orte zum andern, alle Heimlichkeiten offenbahren und in kurtzer Zeit zu erkennen geben kann. Die andere: Wasser-Harnisch, vermöge dessen jemand etliche Stunden, ohne Gefahr.... unter Wasser seyn und sein Vorhaben verrichten kan. Die dritte: Lufft Hosen, mit welchen man ohne Gefahr über die grösten Wasser gehen kan. Der vierdte: Schwimm-Gürtel. Franckfurt, 1722. W. 4 remarkable pl. sm. 8vo. boards. 110.—
    The first work which treats of a scaphander (Wasser-Harnisch). Very remarkable and scarce.

Prices are in Dutch guilders

511. **Burnley, J.**, The history of wool and woolcombing. London, 1889.
Av. de nombr. portr., pl. et ill. toile.   7.50

512. **Casati, P.**, Fabrica et uso del compasso di proportione. Bologna,
1664. Av. pl. de figg. et figg. dans le texte. 4to. vél. souple. 18.—
Quelques pages très légèr. tachées d'eau.

513. — Même ouvrage. Bologna, 1685. W. 4 pl. 4to. boards, uncut.
30.—

514. **(Ceredi, Gius.)**, Tre discorsi sopra il modo d'alzar acque da luoghi bassi. Per adacquaer terreni. Per levar l'acque sorgenti dalle
campagne. Per mandare l'acqua de bere alle città, che n'hanno
bisogno. Parma, S. Viotti, 1567. W. 2 (in stead of 4) pl. and woodcuts. 4to. calf.   10.—

515. **Cessart, L. A. de**, Description des travaux hydrauliques. Paris,
1806, 08. 2 vol. Av. portrait et 67 pl. gr. in-4to. cart., n. r. 15.—

516. **Coets, H.**, Beschrijvinge van vlakke sonnewijsers. Leiden, 1705.
Av. pl. 4to. vél.   5.—

517. **Conférences** du Palais du Trocadero (à l'occasion de l'Exposition
univ. internat. de 1878, à Paris). Comptes rendus sténograph.
Paris, 1879. 3 vol.   7.50
I. Industrie-Chemins de fer. — Travaux publics-Agriculture. — II. Arts-
Sciences. — III. Enseignement. — Sciences économiques-Hygiène.

518. **Dekker, P. M.**, Ontwikkeling van het baggermateriaal. 's-Grav.
1925. Av. pl. et ill. 4to. toile. (15.—)   7.50

519. — Dredging and dredging appliances. W. introd. by R. Madfield.
London, 1927. Av. pl. et ill. 4to. toile. (21.60)   7.50

520. **Demachy**, De sterkwaterstooker, zoutzuur- en vitrioolöliebere-
der. Dordrecht, 1788. W. 10 pl.   5.—
Volledige beschrijving van alle konsten, ambachten, enz. II.

521. **Dou, J. P.**, Tractaet vant maken ende gebruycken eens nieu ghe-
ordonn. mathemat. instruments. Amst., W. Janssen, 1620. Av.
figg. pet. in-8vo.   15.—
L'auteur était un des premiers géomètres des Pays-Bas.

522. **Dugdale, W.**, The history of imbanking and draining of divers
fens and marshes, both in foreign parts and this kingdom, and of
the improvements thereby. Extracted from records, mss. a.o.
authentic testimonies. 2d ed. by Ch. N. Cole. London, 1772. Av. 11
cartes. fol. d. veau. (Rel. cassée).   25.—
Les chap. I-X de cet ouvrage recherché traitent l'histoire de l'endigue-
ment dans l'Egypte, la Babylonie, ...la Belgique, la Hollande, ...
l'Amérique, les chap. XI-XXXII celle de l'Angleterre. — Chap. XXXIV.
How Holland and marshland were first gained from the sea. — etc.

522a. **Duverger, J.**, De Brusselsche steenbickeleren, beeldhouwers, bouw-
meesters, metselaars enz. der XIVe en XVe eeuw. M. aanh. over
Klaas Sluter en zijn Brusselsche medewerkers te Dijon. Gent, 1933.
4.—

523. **Ehrmann, F. L.**, Montgolfier'sche Luftkörper oder aerostat.
Maschinen.... Nebst Beschreib. der zwo ersten Reisen durch die
Luft, und D. Würtz Gedanken über die Ursachen des Steigens
dieser Luftkugeln. Strassburg, 1784. Av. 2 pl.   7.50
Les pp. 77/78 restaurées, avec perte de texte.

524. **Faujas de Saint-Fond,** Description des expériences de la machine
aérostatique de Montgolfier. Paris, 1783. Av. 9 pl. veau.   36.—
525. — Même ouvrage. Brux. 1784. Av. 9 pl. cart.   15.—
Réimpression.
526. — Beschrijving der proefneemingen met konstige lugtbollen, mees-
tendeels te Paris opgelaten. Vert. m. aanteek. d. M. Houttuyn. M.
Vervolg. Amst. 1784. 2 vol. W. 9 pl. boards.   12.50
527. **Fontana, C.,** Dell'acque correnti. Le misure, ed esperienze di esse.
I guochi, e scherzi, li quali per mezzo dell'aria, e del' fuoco, ven-
gono operati dall'acque. Con div. ammaestram, al modo di far
condotti, fistole, bottini, e.a. per condurze l'acque ne' luoghi
destinati, etc. Roma, 1696. Av. de nombr. ill., e.a. plus. de fon-
taines. fol. cart.   18.—
528. **Fougeroux de Bondaroy,** Art du tonnelier. Paris, 1763. Av. 148
figg. sur 6 pl. gr. in-fol. cart., n. r.   7.50
A la fin un vocabulaire des ,,termes propres à l'art du tonnelier."
529. **Fouque, V.,** La vérité sur l'invention de la photographie. N. Niépce,
sa vie, ses travaux. Paris, 1867. Av. portrait et facs. *Rare.* 15.—
530. **Franco, I.,** et **P. Labryn,** Locomotives et automotrices à moteurs à
combustion interne. Trad. des textes hollandais et anglais par A.
M. Hug. La Haye, 1932. Av. 185 figg. toile. *Epuisé.*   6.—
531. **Gallon,** Machines et inventions approuvées par l'Academie Royale
des Sciences, depuis son établissement jusqu'à présent; av. leur
description. Paris,'1735. T. I–VI. En 2 vol. Av. 429 pl. en 2 vol.
Ens. 4 vol. 4to. cart.   40.—
Traitent des inventions techniques depuis 1666 jusqu'à 1734. On y
trouve d'innombrables sortes de machines inventées dans ces années,
p. e.: Huyghens, Machine pour mesurer la force mouvante; Machines
hydraul. etc. p. Perrault; Maniere de relever les vaisseaux submerges
p. de Redingues; Plusieurs traités sur des machines pour tirer les vaisseaux à terre; Parasols p. Marius; Clavecin inventé p. Cuisinié; Plusieurs
traités sur de nouvelles pendules; etc.
L'ouvrage est complet avec un 7e vol. paru en 1777.
532. **Geuns, P.,** Kurze Abhandlung wie alle Magneten zu verfertigen
seyn.... Nebst Unterweisung über die Abwechselung der Zähne in
Räderwerke, etc. Zum Dienste der Uhrmacher. S. l. 1769. Av. 6 pl.
pet. in-8vo. veau.   7.50
533. **Graffenriedt, H. K. von,** Compendium sciotericorum, d.i.: Beschreib.
der Sonnen-Uhren wie die mit unverzucktem Circkel, mit sampt
den zwölff himmelischen zeichen und zu Nacht dienende Mond-
Uhrlein sollen auffgerissen werden. Bern, J. Stuber, 1629. Av. de
nombr. figg. — **Born, Z.,** Sieben Sonnenzeiger. Darinnen nebenst
Erkundigung der Stunden gantzer und halber Uhr zuersehen. Lpz.
1633. — In 1 vol. 4to.   15.—
534. **Hasselt, J. van,** en **de Koning,** Bevloeiingen in Noord- en Midden-
Europa. Rapport omtrent irrigatie-inrichtingen. Nijmegen, 1888.
Av. pl. fol. cart.   5.—
535. **(Hébrard, P.),** Caminologie ou traité des cheminées, observat. sur
les différ. causes qui font fumer les cheminées, av. des moyens
pour coriger ce défaut. Dyon, 1756. Av. 21 pl. pet. in-8vo. d. bas.,
n.r.   7.50

Prices are in Dutch guilders

536. **Hermbstädt, S. F.**, Chemische Grundsätze der Kunst Bier zu brauen. 2e verm. Aufl. Berlin, 1819. Av. 3 pl. d. veau.   5.—

537. **Home, Ensayo s.** el blanqueo de los lienzos, segun se practica en Irlanda, Escocia, varios metodos de conocer las aguas gordas, etc. Trad. (d.) Ingles por la version Francesa p. M. G. Suarez. Madrid, 1779. 4to. calf.   15.—
  Treatise on the bleaching of linen-clothes.

538. **Hooper, W.**, Rational recreations, in which the principles of numbers and natural philosophy are elucidated by a series of experiments. Among which all those with the cards. 2d ed. London, 1782–83. 4 vol. Av. 65 pl. pliées. veau.   15.—
  Contient e.a.: A carriage to go without any force but what it receives from the passengers (av. 1 pl.). — To sale by land against the wind (av. 1 pl.). — The camera obscura. — Electricity (av. pl.). — The planetarium. —

539. **Horst, T. van der,** Theatrum machinarum universale, of nieuwe algemeene bouwkunde, waar in.... wordt voorgestelt en geleert, het maaken van veelerley soorten van trappen, met derselver gronden en opstallen. Amst. 1739. 4to. Av. atlas de 30 pl. par J. Schenk. fol. d. veau.   35.—
  Le seul ouvrage contemporain donnant tous les détails d'après lesquels les escaliers furent construits comme on les trouve encore dans les maisons patriciennes des Pays-Bas des 17e et 18e siècles.

540. — **en J. Polley,** Theatrum machinarum universale; of keurige verzameling van.... waterwerken, schutsluizen, waterkeeringen, ophaal- en draaibruggen. Met harde gronden, opstallen en doorgesnedens. Amst. 1757, 37. 2 tom. 1 vol. Av. 2 armoiries, 49 pl. et 6 pl. suppl. gr. in-fol. d. bas.   30.—

541. **Houcke, A. van,** Vak- en kunstwoorden. VI. Ambacht van den loodgieter en zinkbewerker. Gent, 1901, 02. 2 vol. Av. de nombr. figg. (Vl. Akad.). *Epuisé.*   10.—
  Donne les mots français aussi et souvent les mots allemands et anglais.

542. **Hulsius, L.**, Erster (bis vierdter) Tractat der mechanischen Instrumenten. Franckfurt, 1604–05. 4 tom. 1 vol. Av. 23 pl. hors texte et plus. figg. dans le texte. 4to. vél.   30.—
  Contient: I. Bericht desz newen geometr. gruntreiss. Instruments, Planimetria genandt, mit seinen Inductorio und Ramen. — II. Unterricht desz newen Buchsen Quadrants, wie derselbe, das grosse Geschütz zurichten, gebraucht soll werden. — III. Beschreibung desz Jobst Burgi Proportional-Circkels. — IV. Beschreibung desz Instruments Viatorii oder Wegzählers, zu wissen, wie weit man gegangen sey.
  Première édition. On trouve rarement les quatre parties ensemble.

543. **Inleiding** tot de geschiedenis van het automobilisme. Haarlem, 1910. Av. ill. 4to.   1.—

544. **[Katalog] Aeronautik.** Die Sammlungen E. von Sigmundt, Triest, Dr. O. Nirenstein, Wien. Bücher, Kupferstiche und Lithographien, Autographen, Karikaturen, Medaillen, etc. Versteigerung. Luzern, H. Gilhofer und H. Ranschburg, 1934. Av. 30 pl.   2.50

545. **Keirsbilck, J. en V. van,** Vak- en kunstwoorden. V. Ambacht van den metselaar. Gent, 1899. Av. de nombr. ill. (Vl. Akad.). *Epuisé.* 5.—
  Donne les mots français aussi et souvent les mots allemands et anglais.

546. **La Hire, De,** La gnomonique ou methodes universelles, pour tracer des horloges solaires ou cadrans. Paris, 1698. W. 10 pl. 12mo. calf. (Back slightly damaged).   7.50

547. **Le Conte, P.**, La fabrique et l'usage du radiometre, instrument geometrique, et astronomique, utile tant en la mer qu'en la terre. Paris, 1605. W. 1 pl. and figg. 4to. limp vellum. (Last leaf pasted on cover).   125.—

548. **Leuchs, J. Ch.**, Traité complet des propriétés de la préparation et de l'emploi des matières tinctoriales et des couleurs. Trad. de l'allem. Revu pour la partie chimique p. E. Peclet. Paris, 1829. 2 vol. Av. pl. d. veau.   7.50

549. **Leupold, J.**, Theatrum machinarum generale. Schau-Platz mechanischer Wissenschaften. Lpz. 1724–35. 9 vol. Av. 492 pl. fol. cart. orig.   150.—
    Encyclopédie technique allemande la plus étendue du 18e siècle.
    I. (Allgemeine) Mechanik. (Perpetua mobilia. Zahnradgetriebe. Hebezeug. Wärmekraftmaschinen). Av. 71 pl. — II. Wasser-, Bau-Kunst. (Dammbauten. Suchen von Quellen, etc.). Av. 51 pl. — III, IV. Wasser-Künste. (Fontänen, Pumpen, Pumpwerk in Marly, das Versailles und Trianon mit Wasser für die Fontänen versorgt, etc.). 2 vol. Av. 53 et 54 pl. — V. Hebzeuge. Av. 56 pl. (Wagen, Krane, Fahrstühle, Krankenfahrzeuge, etc.). — VI. Gewicht-Kunst und Waagen. (Wagen für Wasser und andere Flüssigkeiten, Barometer, Manometer, Wettergläser, etc.). Av. 57 pl. — VII. Brücken-Bau. Av. 60 pl. — VIII. Rechen und Mesz-Kunst. Av. 45 pl. — IX. Mühlen-Bau-Kunst. 3 tom. 1 vol. Av. 43 pl. Le suppl. du t. VIII manque.

550. **Ludewig, P.**, Die drahtlose Telegraphie im Dienste der Luftfahrt. Berlin, 1914. Av. ill.   1.75

    Luyken. — S e e n r s. 592–594.

551. **Macquer**, De zijdeverwer. N. h. Fr. verm. m. aanmerk. en aanh. d. P. J. Kastelijn. Dordrecht, 1791. Av. 8 pl.   5.—

552. **Maignan, E.**, Perspectiva horaria s. de horographia gnomonica tum theoretica, tum practica ll. IV. Romae, 1648. W. 20 pl. and figg. fol. vellum.   60.—
    „Traité de catoptrique très remarquable pour l'époque où il a paru. On y trouve la méthode de polir les verres pour les lunettes astronomiques; talent que le P. Maignan possédait à un degré peu commun." (Biogr. Univ.).

553. **Meyer, C.**, L'arte de restituire à Roma la tralasciata navigatione del suo Tevere. Divisa in 3 parti. I. G l'impedimenti, che sono nell' alveo del Tevere da Roma à Perugia, e suoi remedij. II. Le difficoltà, che sono nella navigatione del Tevere da Roma sino al mare, et suoi rimedij. III. Nella quale si discorre perche Roma è stata fabricata, e mantenuta sù le sponde del Tevere e si tratta d'alcun'altre propositioni prosicue per lo stato ecclesiastico. Roma, 1685. W. front. and 68 engravings after drawings by C. Meyer. fol. vellum.   75.—

554. — Dell'arte di rendere i fiumi navigabili in varij modi, con altre nuove inventioni, e varij altri segreti. Divisa in 3 parti. Roma, 1696. 3 vols. in 1. W. 103 engravings after C. Meyer. fol. limp vellum. *Fine copy.*   90.—
    I. Diversi lavori d'acqua, molini, machine per rimediar all' innondazione de' fiumi, etc. W. 66 engravings of hydraulic engines, boats (a.o. „navigatione d'Olanda" representing draw-boats), locks, etc. and some views in Rome. — II. Diversi segreti, como il conoscere la valuta dei metalli, carri che caminano senza cavalli (automobiles), differ. forme di calessi, occhiali, etc. W. 25 engravings represent. glasses, a sort of automobile, different carriages, people bearing spectacles, violin players, etc. — III. Osservazioni delle comete che douranno seguire, e dell' ecclisse del primo satellite di

Giove. W. 12 engravings a.o. a portrait of the author and views of buildings in Rome.
This edition contains nearly all the engravings publ. by the Dutch engineer C. Meyer in his books, without text. It is complete and extremely scarce. Engraving no. 67 of the first part has not appeared, engrav. no. 15 of the second part is to be found as engrav. no. 66 in the first one. The indexes do not quite correspond.
An exact collation of the works of C. Meyer is nearly impossible. The paginations and the signatures are nearly always missing, whereas the numbers of the engravings are often changed.

555. **Moedebeck, H.,** Handbuch der Luftschiffahrt, mit besond. Berücksicht. ihrer militar. Verwendung. Lpz. 1886. 2 tom. 1 vol. Av. 4 pl. et figg.   1.75

556. **Natrus, L. van, J. Polly en C. van Vuuren,** Groot volkomen moolenboek; of naauwkeurig ontwerp van allerhande soorten van moolens, met haare gronden en opstallen. Amst., J. Covens en C. Mortier, 1734, 36. 2 vols. in 1. W. 54 pl., engraved by J. Punt. fol. hfcalf. (Mod. binding, slightly damaged).   75.—
The standardwork on Dutch wind-mills, scarce.
Bound up with: **Linperch, P.,** Architectura mechanica of moole-boek van eenige opstallen van moolens, nevens hare gronden. Amst., J. Covens en C. Mortier, 1727. W. 31 (in stead of 32) pl.

557. **Perez Quintana, J.,** Explicacion de las maquinas e instrumentos de que se compone una fabrica para telillas angostas de lana, su coste, el de sus labores, y utilidades que lograra el estado en su establecimiento. Sevilla, 1785. W. 10 folding pl. 4to.   20.—

558. **Perronet, J. R.,** Description des projets et de la construction des ponts de Neuilly, de Mantes, d'Orléans, e.a., du projet du Canal de Bourgogne et de celui de la conduite des eaux de l'Yvette et de Bièvre à Paris. Av. Suppl. Paris, 1782–89. 3 vol. Av. beau portrait de l'auteur p. A. de St.-Aubin d'après C. N. Cochin fils et 74 belles pl., grav. en taille-douce p. J. F. Germain, P. G. Berthault e.a., de ponts et de leurs subdivisions et des instruments, usités pour leur construction. gr. in-fol. d. veau et veau.   60.—
Ouvrage très estimé et fort bien exécuté.
Parmi les planches on trouve e.a. de belles vues sur les ponts de Neuilly, de Mantes, Sainte-Maxence, Orléans, etc., en même temps intéressantes pour la topographie.

559. **Petit, R.,** Comment on construit un aéroplane. Paris, 1909. Av. ill. pet. in-8vo.   1.—

560. **Pfauth, H.,** Neuestes illustr. Taschenbuch der Bayerischen Bierbrauerei. M. Berücksicht. der wichtigsten theoret. Sätze. Stuttgart, 1870. Av. 11 pl. cart.   0.75

561. **Polak, J.,** Zakboekje en volledige instructie voor suikerkokers en distillateurs, alsmede het aanzetten en distilleeren van rum. Paramaribo, 1863. pet. in-8vo.   1.50

562. **Reaumur, de,** L'art de convertir le fer forgé en acier, et l'art d'adoucir le fer fondu. Paris, 1722. Av. 17 pl. de figg. 4to. veau, dos doré. *Bel ex.*   24.—

563. **Résal, H.,** Traité de cinématique pure. Paris, 1862.   2.—

564. **Richardson, J.,** Vorschläge zu neuen Vortheilen beym Bierbrauen. Nebst Beschreibung seines neuerfundenen Instruments, um den Gehalt des Bieres zu erforschen. A. d. Engl. von D. L. Crell. Berlin, 1788. Av. 1 pl. d. veau.   5.—

565. **Rohart, F.**, Traité de la fabrication de la bière. Suivi d'un project de brasserie-modèle p. A. Riche. Paris, 1848. 2 vol. Av. 3 pl. et figg. 6.—

566. **Rössling, C. W.**, Der wohlerfahrene Küfer oder Büttner. Ulm, 1838. Av. 14 pl. de figg. 5.—

567. **Sainte Marie Magdeleine, P.**, Traitté d'horlogiographie, conten. plus. manieres de construire toutes sortes de lignes horaires, e.a. cercles de la sphere. Avec qq. instrumens, pour connoître les heures durant la nuit, etc. Avec les deux boussoles, en usage pour la navigation des deux mers, etc. 4e ed. augm. Lyon, 1691. W. front. and 81 pl., contain. a large number of figg. sm. 8vo. hfcalf. 28.—

568. **Schottus S. J., P. C.**, Mechanica hydraulico-pneumatica. Acc. experimentum novum Magdeburgicum. Herbipoli, 1657. W. front., 46 pl., figg. and music. 4to. vellum. 36.—
   Contains a.o.: De organis hydraulicis, aliisque instrumentis hydropneumaticis. — etc.
   Graesse is mistaken in mentioning 56 pl.

569. — Technica curiosa, s. mirabilia artis. Herbipoli, 1664. 2 vol. Av. front. et pl. 4to. veau. 15.—

570. **Schwenter, D.**, Deliciae physico-mathematicae.... darinnen 663 Kunst stücklein, Auffgaben, etc. auf der Rechenkunst, Landtmessen, Naturkündigung, etc. begriffen seindt. Nürnberg, 1636. — **G. Ph. Harsdörffer**, Deliciae mathematicae et physicae 2er (und) 3er Theil, bestehend in (je) 500 Kunstfragen, aus A. Kirchero, M. Mersennio, R. des Cartes, C. Drebbelio, u A. Nurnberg, 1677, 1653. 2 vol. — Ens. 3 vol. Av. titre gravé, 2 front., quelques pl., de nombr. figg., grav. s. bois dans le texte et qq. pp. de musique. 4to. vél. 40.—
   Contient e.a.: Astronomia und Astrologia. — Gnomonica oder Zubereitung der Sonnen- und Schlaguhren. — Motum oder künstlicke Bewegung. — Pyrobolia, und sonsten durchs Feur zu verrichten. — Pneumatica. — Schreibkunst. — Chymia, — etc.
   On y trouve e.a. une belle et curieuse gravure en grandeur de la page, appartenant à l'article: Wie das Schach und Dammspiel mit lebendigen Personen zu spielen? L'ouvrage contient aussi: Discurs von dem Damm und Schachtspiel, et un autre traité sur ce sujet, accompagné d'une fig. d'un damier. Une des pp. de musique contient: Die Stimme der Nachtigal, der Haanen Geschrey, etc.
   D'une des figg. mobiles une partie manque.

571. **Secrets** concernant les arts et métiers. Nouv. éd. augmentée. Brux. 1747. 2 vol. d. veau. 7.50
   Contient des centaines de recettes utiles et peu connues, ainsi des secrets pour colorer l'ébène, les bois, les os, des secrets pour mouler, pour dorer, des secrets pour parfumer les tabacs, etc. Le 2e vol. traite exclusivement de l'art du teinturier.

572. **Serristori, L.**, Sopra le macchine a vapore. Firenze, 1816. Av. 8 pl. se dépliant. d. veau. 48.—
   Pp. 59—97 (fin): Navi a vapore, e loro utilita, specialmente a riguardo del commercio di Italia.
   Opuscule extrêmement rare.

573. **Stempelius (Goudanus), G.**, et **A. Zelstius**, Utriusque astrolabii

Prices are in Dutch guilders

tam particularis quam universalis fabrica et usus. Leodii, 1602.
W. 11 pl. 4to. limp vellum. (16+40+40 +99+12 pp.).   125.—
Stempels or Stempelius was the inventor of a new instrument to measure
the distance of the stars.
Fine copy, with the revolving diagrams all quite intact.

574. **Stofler, J.**, Elucidatio fabricae ususque astrolabii. Oppenheym, J.
Kobel, 1513 (*at the end* 1512). W. title in a border, numer. figg.
and fine initials, white on a black background, all engraved on
wood. fol. hfvellum. (Mod. binding).   180.—
Proctor, nr. 11922. First edition, very scarce, of this work by the famous
mathematician, astronomer and astrologer, of whom a.o. Melanchton was
one of the disciples. It contains many particulars about horoscopes and
astrological predictions.
The beautiful woodcuts have been used as specimens for the illustrations
of several later works on the same subject.

575. **Strada à Rosberg, J. de**, Kunstl. Abrisz, allerhand Wasser-, Wind-,
Rosz- und Handt Mühlen, beneben Pompen, auch andern Machinen,
Brunnen und Wasserwerck. Franckfurt, 1617, 18. 2 tom. 1 vol.
Av. 2 titres dans de beaux encadrements et 100 pl. fol. cart.   30.—
L'explication du t. I manque; le t. II taché d'eau.

576. **Swinden, J. H. van, e.a.**, Inrigting en gebruik der octanten en sex-
tanten van Hadley. 2e dr. Amst. 1826. Av. 2 pl. se dépliant. cart.,
n. r.   7.50

577. **Tissandier, G.**, Les ballons dirigeables. Paris, 1885. Av. pl. — **F.
Ferber**, Les progrès de l'aviation depuis 1891 par le vol plané. Paris,
1905. — Ens. 2 vol.   1.50

578. **Ubaldo, G.**, Mechan. Kunstkammer. I. Von Wag, Hebel, Scheiben,
Haspel, Keil und Schrauffenwerckh. A. Ital. u. Lat. übers. und d.
Addit. erklärt d. D. Mögling. Franckfurt a.M. 1629. Av. titre gravé
et 42 pl. de figg. fol. d. veau.   15.—
Cette édition allemande, dont le t. I seul a paru, contient une traduction
de l'ouvrage de G. Ubaldo, Del monte mechanica, B. Baldi, Problemata
Aristotelis, e.a. écrits.

579. **Valicourt, E. de**, Nouveau manuel de photographie. Paris, 1851.
Av. 1 pl. pet. in-8vo.   1.—

580. **Vidal, L.**, Traité pratique de photoglyptie. Paris, 1881. Av. 1 phot.
et figg.   1.25

581. **Wegner von Dallwitz, R.**, Der praktische Luftschiffer. Rostock,
1909. Av. 42 ill. toile.   1.25

582. **Weidler, J. F.**, Tractatus de machinis hydraulicis foto terrarum
orbe maximis Marlyensi et Londinensi. Vitemb. 1728. Av. 5 pl. —
Id., Institutiones geometriae subterraneae. Ibid. 1726. Av. 4 pl. —
Ens. 2 ouvrages en 1 vol. 4to. cart.   6.—

583. **Weiss, H.**, Grundsätze der Kinematik. Lpz. 1900. Fasc. 1 (seul).
Av. atlas de 10 pl. 4to. (6.—)   3.—

584. **Westen, W. van**, Mathematische vermaecklycheden. 6en dr. Amst.,
M. de Groot, 1673. 3 tom. 1 vol. vél.   12.—
Contient e.a.: Een instrument te maken dat van verre doet hooren. — Van
verscheyden soorten van brillen. — Van de magneet-steen, ende van de
zeylnaelden. — Van het kaets-spel, kegel-spel, etc. — Van het dammen
ende van het schaeck-spel. — etc. La 3e partie traite de la pyrotechnique.

585. **Wiebeking**, Memoire sur des ponts suspendus en chaines de fer.
rel. aux ponts construits dans le dernier temps en Angleterre et
en Russie. Münich, 1832. Av. 6 pl. 4to. cart. orig.   6.—

586. **Wildeman, J. E.,** Het vormen, bakken en gebruiken van zware tigchelsteenen. Amst. 1825. Av. 3 pl. 1.—

587. **Wood, J. T.,** Das Entkälken und Beizen der Felle und Häute. N. d. Eng. bearb. von J. Jettmar. Brschw. 1914. Av. portr. et 34 figg. toile. (7.20) 2.50

588. **Zahn, J.,** Oculus artificialis teledioptricus, s. telescopius. Ed. 2a. Norimb. 1702. Av. front., tables, pl. et figg. fol. vél. 20.—

589. **Zerella y Ycoga, M. de,** Tratado general y matematico de reloxeria. Acomp. de los elementos necesarios para ella, comoson aritmetica, algebra,.... astronomía, geografía, física, maquinaria, musica y díbuxo. 2a impr. Madrid, 1791. W. 22 pl. sm. fol. calf. (Binding faded). 20.—

590. **Zonca, P.,** Novo teatro di machine et edificii per varie et secure operationi. Padoua, 1621. W. front. and 42 ill. at full-page size. fol. limp vellum. 50.—
   Contains a.o. the description of locks and wind-mills, of the manufacture of oil, of different printer's presses, etc.

591. **Zyl, J. van,** Theatrum machinarum universale, of groot algemeen moolen-boek, behelz. de beschryving en afbeeldingen van allerhande soorten van moolens, derzelver opstallen, en gronden. Amst., P. Schenk, 1761. 2 tom. 1 vol. Av. front. et 62 pl., grav. p. J. Schenk. gr. in-fol. d. bas. 60.—
   Ajouté: Verklaring van een zy-tweyndery. (Amst. 1761). 4 pp. de texte. 4to. Av. 2 pl. gr. in-fol.

592. **Luyken, J.,** Afbeelding der menschelyke bezigheden. Amst., R. en J. Ottens, (v. 1730). Av. 100 grav. en taille-douce. 4to. d. veau. 60.—
   Les jolies gravures sont de la plus haute importance pour la connaissance des divers métiers dans le 17e siècle. Chaque feuillet est entièrement gravé et imprimé d'un côté et a une légende en vers.

593. — Même ouvrage sous le titre: Spiegel van het menselyk bedryf. Amst., P. Arentz en P. van der Sys, 1704. Av. front. et 100 jolies grav. — **Id.,** De zedelyke en stichtelyke gezangen. Ib., id., 1709. Av. front. et grav. dans le texte. — En 1 vol. pet. in-8vo. veau. 25.—
   3e édition de „Het menselyk bedryf".

594. — Même ouvrage. Amst., F. Houttuyn, 1767. Av. front. et 100 jolies grav. pet. in-8vo. br., n. r. 15.—

595. **Wijk, W. E. van,** Le nombre d'or. Etude de chronologie technique. Suivie du texte de la Massa compoti d'Alexandre de Villedieu avec traduction et commentaire. La Haye, 1936. Av. 17 pl. et figg. 10.—

596. — Decimal tables for the reduction of Hindu dates from the data of the Surya-Siddhānta. The Hague, 1938. W. portrait and tables. 4.—

597. — De gregoriaansche kalender. Een technisch-tijdrekenkundige studie. 's-Grav. 1938. W. 12 portr. and pl. 4.—

Prices are in Dutch guilders

SUPPLEMENT

598. **Assemanus, S.**, Globus caelestis Cufico-Arabicus Veliterni Musei Borgiani. Praemissa ejusdem de Arabum astronomia dissert. Patavii, 1790. Av. 3 gr. pl. 4to. veau. (Rel. légèr. endomm.). 35.—

599. **Behm, G.**, Polymetrum, hoc est, novum instrumentum, ad plerasque mathematicas dimensiones rité facileque peragendas, accommod. explicat. Salisburgis, 1682. Av. 1 pl. pet. in-8vo. vél. 3.50
Qq. ff. jaunis, cachets sur le titre.

600. **Belidor,** Nouveau cours de mathématique à l'usage de l'artillerie et du génie. Paris, 1725. Av. 34 pl. 4to. veau. 5.—

601. **Berchuys, C. H. J. van,** De doliometria. Daventriae, 1839. Av. pl. veau. 1.50

602. **Bernoulli, D.**, Hydrodynamica sive de viribus et motibus fluidorum. Argent. 1738. Av. 11 pl. 4to. 2.50

603. **Bézout,** Cours de mathématiques à l'usage du corps royal de l'artillerie. Paris, 1772. 2 vol. Av. 9 pl. veau. 2.50

604. **Biehler, Ch.**, Les développements en séries des fonctions doublement périodiques de 3e espèce. Paris, 1879. 4to. dos et coins veau. 3.50

605. **Blakey,** Observat. sur les pompes à feu avec balancier et sur la nouvelle machine à feu. Av. remarques sur la situation de la Hollande, sur les moyens, dont on se sert pour la rendre habitable etc. S. l. 1777. Av. 3 pl. 4to. 5.—

606. **Blondeel, V. J.**, De pressione fluidorum. Utr. 1743. Av. figg. 4to. 2.—

607. **Bode, J. E.**, Uranographia s. astrorum descriptio XX tabulis aeneis incisa ex recent. astronomorum observation. Berol. 1801. Av. titre gravé et les cartes célestes I et XV—XX. très gr. in-fol. cart. 7.50

608. **Bolyal, J.**, La science absolue de l'espace; suivie de la quadrature géométrique du cercle. (Trad. du hongrois, av. notice biograph. p. F. Schmidt). Paris, 1868. Av. figg. 2.50

609. **Bouguer,** Traité d'optique sur la gradation de la lumière. Publ. p. de la Caille. Paris, 1760. Av. 7 pl. 4to. veau. 7.50

610. **Buch** der Erfindungen, Gewerben und Industrien. Hrsg. von Bierbaum, Böttger, u. A. 7e verb. Aufl. Lpz. 1876-83. 7 vol. Av. de nombr. pl. et figg. d. mar. 5.—

611. **Buchner, J. P.**, Tabula radicum quadratorum et cuborum ad radic. 12000 extensa. Norimb. 1701. Av. front. 12mo. d. veau. 4.50

612. **Burgers, J. M.**, Het atoommodel van Rutherford-Bohr. Haarlem, 1918. 3.—
Tirage à part des Archives du Musée Teyler.

613. **Capella, Mart.**, Opus. De nuptiis philologiae et mercurii ll. II; de grammatica, de dialectica, de rhetorica, d e g e o m e t r i a, d e a r i t h m e t i c a, de a s t r o n o m i a, de musica. (Ed. a J. Chieregato). Mutinae, D. Bertochus, 1500. sm. fol. orig. vellum. 200.—
Hain, nr. 4371. Pellechet, nr. 3225 (showing a slight variation, as there is no engraved border on the titlepage). Proctor, nr. 7215. This is one of the two books Bertochus printed at Modena. All incunabula, printed there, are scarce.
Very good copy, printer's mark on last leaf.

Mat. Nijhoff, The Hague — Cat. No. 632

614. **Catalogus** en korte beschrijving van eene uitgebreide en zeer kostbare verzameling van wis- en natuurkund. werktuigen uitmakende het kabinet van G. van Varik. Vente à Amsterdam, 1825. cart. 5.—
Catalogue de vente d'une précieuse collection de toute sorte d'instruments mathémat., physiques, électriques, nautiques, hydrograph., etc.

615. **Chateleux, P. J. L. de, et J. P. van Rooijen,** Le rapport de Johan de Witt sur le calcul des rentes viagères. Trad. franç. av. commentaire et historique. La Haye, 1937. W. portrait. *Reprint.* 1.—

616. **Crook, W. J.,** Metallurgical spectrum analysis. Stanford University, 1935. Av. 16 figg. et atlas de 24 tabl. 4to. toile. 25.—

617. **Desargues,** Algemeene manier tot de practijck der perspectiven gelijck tot die der meetkunde. Bijeeng. d. A. Bosse. U. h. Fr. d. J. Bara. Amst. 1664. Av. front. et 156 pl. pet. in-8vo. vél. 8.50
Dans la même reliure: **A. Bosse,** Middel tot de practijck der doorzightkunde op tafereelen of regel-lose buytengedaenten. Amst. 1664. Av. front. et 32 pl.

618. **Diepenbeek, Abr. van,** De wiskonstige reken-konst. Amst. 1711. T. I. pet. in-8vo. vél. 2.—

619. **Dijkmans, C. J. M., en B. J. van der Ploeg,** Premieberekening voor levensverzekering met grondgetallen, bewerkt naar de sterftetafels van van Pesch. Amst. 1889. gr. in-fol. d. veau. 5.—

620. **Eckhardt, A. G.,** Description d'un graphomètre universel, nouv. instrument, propre à dessiner toutes sortes d'objets. La Haye, 1778. T. 1 (seul paru). Av. 12 pl. gr. in-fol. 6.—

621. — Beschrijving van een algem. graphometer, zijnde eeñ allernauwkeurigst teken-werktuig. 's-Grav. 1778. T. I (seul paru). Av. 12 pl. gr. in-fol. 4.50

622. **Engelman, J.,** Het regt gebruik der natuurbeschouwingen, geschetst in eene verhandeling over de sneeuw figuuren. Haerlem, I. v. d. Vinne, 1747. Av. 29 pl. vél. 4.50

623. — Même ouvrage. 2e uitg. met voorrede van P. Boddaert. Utrecht, 1771. Av. 29 pl. cart. 4.50

624. **Euclides,** Elementarum VI ll. priores. Opera H. Coetsii. Ed. 2a. Amst., H. et Vid., Th. Boom, 1705. Av. front. et figg. pet. in-8vo. vél. 2.50

625. — De ses eerste boeken der beginselen van Euclides, gedemonstr. d. H. Coets. Leyden, 1715. Av. figg. pet. in-8vo. vél. 4.50

626. — Elementarum ll. VI ac XI et XII. C. explic. Chr. Clavii ed. a J. H. v. Lom. Amst., H. Vieroot, 1738. Av. 36 pl. de figg. vél. 4.50

627. **Eversdijck, C. F.,** Tafelen van interest. Middelburg, J. Fierens, 1663. 4to. vél. 10.—

628. **Franco, I., and P. Labrijn,** Internal-combustion. Locomotives and motor coaches. The Hague, 1931. Av. 185 figg. toile. *Epuisé.* 6.—
„.... a book that could be consulted by anyone interested in the problem of the application of combustion engines on railways in order to see what had already been attained in this direction in the various countries of the world".

629. — — Verbrennungs- Motor-Lokomotiven und Triebwagen. Haag, 1932. Av. 185 figg. toile. 6.—

630. **Frisius, G.,** Anweisung zur Physica. Lpz. 1696. 7.50
Curieux.

Prices are in Dutch guilders

631. **Fullone, A.**, Descrittione et uso dell' holometro per saper misurare tutte le cose. Venetia, G. Ziletti, 1564. — **Belli, S.**, Libro del misurar con la vista. Venetia, id., 1566. — En 1 vol. Av. figg. 4to. cart. 12.—
Ecritures sur la dernière page.

632. **Fuss, N.**, Instruction détaillée pour porter les lunettes au plus haut degré de perfection, tirée d'Euler, avec la description d'un microscope. St. Petersbourg, 1774. Av. 2 pl. 4to. d. veau. 5.—

633. **Gilbert, Ph.**, Cours d'analyse infinitésimale. Partie élémentaire. 4e éd. Paris, 1892. 5.—

634. **Goldman, N.**, Tractatus de usu proportionatorii s. circini proportion. Anleitung vom Gebrauch des Ebenpassers oder Proportionalcirkels. (Texte latin et allemand). Lugd. Bat., Ph. de Croy, 1656. Av. 16 pl. fol. vél. 9.—
Timbre sur le titre.

635. **Goldstein, L.**, Les théorèmes de conservation dans la théorie des chocs électroniques. Paris, 1933. cart. 1.—
Actualités scientifiques et industrielles. No. 70.

636. **Graaf, A. de**, Inleyding tot de wiskunst; of de beginselen van de geometrie en algebra. 2e dr. Amst. 1706. Av. figg. 4to. vél. 5.—

637. —— De vervulling van de geometrie en algebra begrepen in de Inleyding tot de wiskunst. Amst. 1708. Av. figg. 4to. 3.50

638. **Hermannus, J.**, Phoronomia s. de viribus et motibus corporum solidorum et fluidorum. Amst. 1716. Av. front. et 12 pl. 4to. 2.50

639. **Hertel, Chr. G.**, Vollständige Anweisung zum Glass-Schleiffen, wie auch zu Verfertigung derer optischen Machinen die aus geschl. Gläsern zubereitet werden. Vorrede von C. Wolffes. Halle, 1716. Av. front. et 19 (sur 20) pl. d. vél. 7.50
Pl. 18 manque, qq. ff. lim. ont souffert de l'eau.

640. **Hocker, J. L.**, Einleitung zur Erkenntnis und Gebrauch der Erd- und Himmels-Kugel. Nürnberg, 1734. Av. front. et 10 pl. color. 4to. d. vél. 24.—

641. **Hoek, M.**, De kometen van 1556, 1264 en 975, en hare vermeende identiteit. 's-Grav. 1857. Av. 1 pl. 4to. 2.50

642. **Horst, J. D.**, Physica Hippocratea, Tackenii, Helmontii, Cartesii, Espagnet, Boylei, etc. aliorumque recentiorum commentis illustr. Francof. 1682. pet. in-8vo. 3.—
Timbre sur le titre.

643. **Horst, T. van der, en J. Polley**, Theatrum machinarum universale; of keurige verzameling van.... waterwerken, schutsluysen, waterkeeringen, enz. Met haare gronden, opstallen en doorgesneden. Amst., W. Holtrop en N. T. Gravius, s. d. en Amst., P. Schenk, 1774. 2 tom. 1 vol. Av. 49 pl. et 6 pl. suppl. p. J. Schenk. gr. in-fol. d. veau. 20.—

644. **Hues, R.**, Tractatus de globis, coelesti et terrestri eorumque usu. Semelque atque iterum à J. Hondio excusus, et nunc elegantibus iconibus et figuris locupletatus, ac de novo recognitus opera ac studio J. I. Pontani. Amst., H. Hondius, 1624. Av. 1 pl. et plus. figg., grav. s. bois. 4to. cart. 48.—
Un des ouvrages les plus importants sur les globes anciens.
Nom et timbre sur le titre.

Mart. Nijhoff, The Hague — Cat. No. 632

645. **Jullien, (M.)**, Problèmes de mécanique rationnelle. 2e éd. Paris, 1866, 67. 2 tom. 1 vol. d. veau. *Rare.*   7.50
646. **Jungenickel, A.**, Schlüssel zur Mechanica, d.i. Beschreib. der vier Haupt Instrumenten der Machination, als desz Hebels, Betriebs, Schrauben, Kloben. An den Tag gegeben d. M. Stier. Nürnberg, (1661). Av. front. et 137 figg. 4to. d. vél.   9.—
647. **Keill, J.**, Introductiones ad veram physicam et veram astronomiam quibus accedunt trigonometria etc. L. B., J. et H. Verbeek, 1725. Av. pl. 4to. vél. cordé.   7.50
648. **Knabbe, W.** von, Fraiser und deren Rolle bei dem derseitigen Stande des Maschinenbaues. Charkow, 1893. 8vo. Av. atlas de 452 figg. sur 39 pl. fol.-obl.   6.—
649. **Krazer, A.**, Lehrbuch der Thetafunktionen. Lpz. 1903. toile.   10.—
650. **La Caille, D. de**, Lectiones elementaris opticae. Ex ed. Paris. in Lat. trad. Acc. Brevis theoria micrometri objectivi a R. J. Boscovich. Viennae, 1750. Av. pl. 4to. d. veau.   6.50
651. — Lectiones element. mathematicae. In lat. trad. Viennae, 1758.
    — **Id.**, Lect. elem. astronomiae, geometriae et physicae. Viennae, 1757. Av. 9 pl. — Ens. 2 tom. 1 vol. 4to. d. veau.   5.—
652. **Lagrange, J. L.**, De la résolution des équations numériques de tous les degrés. 3e éd. Paris, 1826. 4to. veau.   6.—
    Edition la plus estimée.
653. **Le Clerc**, Thesaurus geometriae practicae. London, 1737. Av. de nombr. pl. pet. in-8vo. veau.   2.50
654. — Même ouvrage. Ultr. 1748. Av. grav. pet. in-8vo.   2.50
655. **Leeuwen, H. J. van**, Vraagstukken uit de electronentheorie van het magnetisme. Leiden, 1919.   2.—
656. **Lobatto, R.**, Beschouwing van den aard, de voordeelen en de inrichting der maatschappijen van levensverzekering bevatt. eene verklaring der ware gronden van berekening tot het ontwerpen van duurzame weduwen-fondsen. Amst. 1830. *Rare.*   5.—
657. **Lorentz, H. A.**, Lessen over theoretische natuurkunde. Bew. d. A. D. Fokker, G. L. de Haas-Lorentz, H. Bremekamp, e.a. Leiden, 1919—25. 8 vol. Av. figg. d. rel. (1 br.). (35.—)   10.—
    I. Stralingstheorie. — II. Theorie der quanta. — III. Aethertheorieën en aethermodellen. — IV. Thermodynamica. — V. Kinetische problemen. — VI. Het relativiteitsbeginsel voor eenparige translaties. — VII. Entropie en waarschijnlijkheid. — VIII. De theorie van Maxwell.
658. **Maginus, J. A.**, De planis triangulis l. I. Ejusdem de dimetiendi ratione per quadrantem, et geometricum quadratum ll. V. Venetia, J. B. Ciottus, 1592. Av. figg. 4to. vél.   36.—
    Edition originale.
659. **Michel, H.**, Introduction à l'étude d'une collection d'instruments anciens de mathématiques. Anvers, 1939. Av. portrait et 14 pl. 4to.   6.—
    L'école brabançonne. — Erasme Habermel. — Le siècle de Shah Abbas. — Le Tum-Fum-La-Fou-Seu-Chim-Tam. — La révocation de l'édit de Nantes. — Description technique des anciens instruments de mathématiques. — Bibliographie. — etc.
    Tirage restreint.
660. **Moore, Th. W.**, Logarithms of numbers chiefly to six places. Bloomington Ind., 1932. 8vo-allongé. toile.   1.50

661. **Morgenster, J.**, Werkdadige meetkonst.... hoe dat al 't gene een ingenieur en landmeter te meten voorvallen kan.... (M.) verhandel. van roeden en landmaten etc. Zwolle, 1703. pet. in-8vo. vél. 6.—

662. — Même ouvrage. Verm. d. J. H. Knoop. 2e dr. 's-Grav. 1757. Av. 30 pl. et cartes et grav. dans le texte. d. veau. 7.50

663. **Nispen, M. van**, De beknopte lant-meetkunst, leerende in 't korte alles wat in de practijcke des landt-meters voorkomen kan. 2e dr. (M.) het tractaet van de landmaten d. J. P. Dou ende G. Eversdijck e.a. Dordrecht, 1669. Av. front. (légèr. endomm.) et ill. pet. in-8vo. vél. 7.50

664. — Même ouvrage. Dordrecht, 1768. vél. 7.50

665. **Perrault**, Recueil de plusieurs machines, de nouvelle invention. Paris, 1700. Av. 11 pl. se dépliant de machines. 4to. cart. 7.50

666. **Physique.** — **Ecrits** en néerlandais et en anglais sur les résultats d'expériments faits dans le laboratoire de physique de Leide (par W. J. de Haas, H. Kamerlingh Onnes, W. H. Keesom, G. P. Nijhoff, G. J. Sizoo, A. Th. van Urk, e.a.), celui d'Amsterdam (par P. Geels, A. Michels e.a.) et d'autres écrits sur des sujets analogues (quelques-uns en allemand et en français). 1921—31. Ens. env. 100 pièces. 30.—
    Pour la plupart des tirages à part des „Verslagen" ou „Proceedings" de l'Académie royale des sciences d'Amsterdam", des „Communications from the physical laboratory de Leide", puis du „Zeitschrift f. Physik", de la „Royal Society", etc.

667. **Pickering, E. C., J. C. Kapteyn** and **P. J. van Rhijn**, Durchmusterung of selected areas. Groningen, 1923, 24. 2 vols. roy. 4to. (25.—) 15.—

668. **Pomodoro, G.**, Geometrica pratica ridotta in tavole cinquantuano con le spiegazioni di G. Scala. Roma, 1772. Av. titre gravé, portant la date 1771, et 51 pl. gr. in-fol. 10.—
    Ex. av. 51 pl. au lieu de 50 comme le titre l'annonce.

669. **Puyt, S. J. de**, Grondbeginselen der meetkunde, vervatt. de zes eerste, het elfde en twaalfde boek van Euclides. Leyd. 1784. Av. pl. d. veau. 3.50

670. **Quadri, L.**, Tavole gnomoniche per delineare orologj a sole. Bologna, 1733. Av. 6 pl. de figg. — Tavole per regolare di giorno in giornogli orologj a ruote, con una tavola perpetua del principio dell' aurora, levar del sole, etc. como app. alle gnomoniche di L. Quadri. Bologna, 1736. — 2 tom. 1 vol. 4to. cart. 5.—

671. **Regimento** do estrolabio e do quadrante. Tractado de spera domundo. Hrsg. van Bensaude. *Einleitung.* München, 1914. 4to. (34 pp.). 2.50
    Parut en même temps en français dans la Collection de documents pour l'histoire de la science nautique portugaise, t. I.

672. **Rohault, J.**, Physica. Latinè vertit recens. et adnot. ex I. Newtoni philosophia amplific. et ornavit S. Clarke. Londini, 1718. Av. 26 pl. de figg. veau. 7.50

673. **Romp, H. A.**, Oil burning. The Hague, 1937. W. 7 tables and 262 figg. 4to. cloth. 12.—
    This work deals with the history, present theories and forms and future of oil burning. It contains the description of over 200 burners; numerous patents and other data are quoted.

674. **Scherfer S. J., C.,** Institutiones mechanicae, s. de motu et aequilibrio corporum solidorum et fluidorum. Vindob. 1773. 2 tom. 1 vol.
Av. 20 pl. 4to. d. rel. 7.50

675. **Sems, J.,** ende **J. Pz. Dou,** Practijck des landmetens, leerende alle
rechte ende kromzydige landen, bosschen, etc. meten.... (Van
het ghebruyck der geometr. instrumenten.... Desghelijcks caer-
ten maecken....). Verm. m. hondert geometrische questien met
haer solutien d. S. Hansz. Amst., W. Jansz., (v. 1620). Av. 6 pl. et
figg. dans le texte. veau. (Un plat endomm.). 9.—

676. **Serarius, P.,** Naerder bericht wegens die groote conjunctie ofte
t'samenkomste van allen planeten in het teecken .... de Schutter
te geschieden den 1—11 Dec. 1662. Amst. 1662. 4to. 5.—

677. **Sturm, L. Chr.,** Der wahre Vauban, oder der von den Deutschen
und Holländern verbesserte Französ. Ingenieur, w. i. die Arith-
metik, die Geometrie.... die Kriegsbaukunst, und Fortification.
Av. 22 pl. — **Goulon, von,** Bericht von Belagerung und Vertheidi-
gung einer Vestung. — Nürnberg, 1761. 2 tom. 1 vol. 4to. veau. 7.50

678. **Thomson, G. P.,** Applied aerodynamics. London, 1920. Av. pl. 4to.
toile. (25.—) 12.50

679. **Tutton, A. E. H.,** Crystallography and practical crystal measure-
ment. London, 1911. Av. ill. toile. (18.—) 12.—

680. **Uhlenbeck, G. E.,** Over statistische methoden in de theorie der
quanta. 's-Grav. 1927. 2.—

681. **Valentin Mennher de Kempten,** Arithmetique seconde. Anvers, Ian
Loë, 1556. Av. portrait et de nombr. jolies grav. s. bois petites et
grandes, figg. mathématiques, paysages, métiers divers, etc., dont
qq.-unes signées Ant. Sylvius. pet. in-8vo. veau. 350.—
   Première édition, infiniment rare, divisée en trois parties: De la signification et prononciation des ciffres. — La regle d'algebre. La geometrie.
   La première partie est d'importance pour la tenue des livres.

682. **Vlacq, A. A.,** Tabulae sinuum, tangentium; et secantium, et loga-
rithmi sinuum etc. ab unitate ad 10000. Ed. aucta. Amst., J. Boom,
1921. pet. in-8vo. d. veau, n. r. 3.50

683. — Même ouvrage. Amst. 1742. pet. in-8vo. d. veau. 2.50

684. **Waals, Van der,** centenary number. The Hague, 1937. W. portr. 5.—
   By far the greater part of the articles is in English, the others in German
   or French.
   Physica. Vol. IV, nr. 10.

685. **Warin, A.,** Patallig of twaalftallig stelsel. 's -Grav. 1842. 1.—

686. **Warius, P.,** Nieuwe verklaring over de proportionaal passer. Amst.,
J. Loots, 1708. Av. 13 pl. 4to. cart. 7.50
   A la fin: Beschrijving der Telegraphe, d. E. A. Kieser. 2 pp. en ms. Av. 1
   dessin.

687. — De zes eerste, elfde en twaalfde boeken Euclidis, verton. de voornaamste gronden, der meetkonst op een korte manier gedemon-
streert. Amst. 1717. 2e dr. Av. pl. pet. in-8vo. vél. 2.50

688. **Wiaerda, H.,** Naauwk. verhandelinge van de eerste uitvindingen
en uitvinders. Amst. 1733. Av. front. pet. in-8vo. vél. 4.50

689. **Eymers, J. G.,** Fundamental principles for the illumination of a
picture gallery. Tog. with their application to the illumination of
the Municipal Museum at the Hague. W. pref. by L. S. Ornstein.
The Hague, 1936. W. 35 figg. 2.—

Prices are in Dutch guilders

MARTINUS NIJHOFF — PUBLISHER — THE HAGUE

VAN DER WAALS
CENTENARY NUMBER

25 papers on physical subjects, contributed by 37 Dutch and foreign physicists in honour of Dr. J. D. van der Waals

(Physica, Vol. IV, No. 10, Nov. 1937)

268 pp. with a portrait and 57 illustr. roy. 8vo.   5 guilders

CONTAINS a.o.: R. C. L. BOSWORTH and E. K. RIDEAL, Intermolecular forces in two dimensional systems. — W. KAST u. W. MAIER, Die Wechselwirkung der Moleküle in den anisotropen Flüssigkeiten. — G. SCHMIDT and W. H. KEESOM, New measurements of liquid helium temperatures. I. The boiling point of helium. II. The vapour pressure curve of liquid helium. — H. C. HAMAKER, The London-van der Waals attraction between spherical particles. — J. ERRERA, Examen spectrographique infrarouge des liaisons intermoleculaires. — etc., etc.

Pieter Zeeman
1865 — 25 Mei — 1935
VERHANDELINGEN
OP 25 MEI 1935 AANGEBODEN AAN
PROF. DR. P. ZEEMAN

51 papers on physical subjects, presented to Dr. P. Zeeman, Professor of Physics at the University of Amsterdam on the occasion of his 70th anniversary, by Dutch and foreign professors of physics.

1935. VIII and 424 pp. with a portrait and 4 plates. roy. 8vo.
3 guilders

E. Amaldi und E. Segré, Einige spektroskop. Eigenschaften hochangeregter Atome. — G. E. Hale, The magnetic periodicity of sun-spots. — N. Bohr, Zeeman effect and theory of atomic constitution. — G. E. Uhlenbeck and S. Goudsmit, Statistical energy distributions for a smaller number of particles. — and contributions by T. Mishima and H. Nagaoka, J. Becquerel, E. Cohen, H. A. Kramers a.o.

MARTINUS NIJHOFF — ÉDITEUR — LA HAYE

Le tome premier vient de paraître de:

JOURNAL

tenu par

ISAAC BEECKMAN

de 1604 à 1634

publié avec une introduction et des notes

par

C. DE WAARD

Isaac Beeckman, ami de Descartes et comme lui s'intéressant aux mêmes questions de physique (prise au sens le plus large), peut être nommé avec honneur parmi les cultivateurs de la science, non seulement parmi ceux des Pays-Bas, mais encore ceux d'autres pays. Il résulte dès maintenant de diverses publications que c'était lui qui a connu et appliqué, comme un des premiers le principe d'inertie; qu'il a participé considérablement à la découverte de la loi de la chute des graves en l'exposant dans ses notes bien avant que Galilée publiât ses premières recherches sur cette loi, et qu'il a [été le premier à formuler les lois du choc des corps mous, se fondant sur le principe de la conservation de la quantité de mouvement. D'ailleurs Beeckman est partisan de la théorie atomistique, à son époque généralement rejetée, et c'était cette hypothèse qui l'amena à admettre la pression de l'air se propageant dans tous les sens, longtemps avant les expériences de Torricelli, expériences dont la signification fut interprêtée faussement même encore plus tard. A l'opinion courante s'oppose aussi la conviction de Beeckman que la lumière se propagerait par une vitesse finie, ce qui ne fut confirmé que par les expériences de Römer en 1677, etc.

Non sans raison on a nommé Beeckman „un des esprits les plus importants parmi les physiciens hollandais" et une figure intéressante et considérable".

On donnera maintenant la publication intégrale de son Journal. Elle comprendra quatre volumes d'environ 365 pages chacun, en format in-4to. Le texte est pour la majeure partie en latin, accompagné de plusieurs facsimilés des dessins originaux; l'introduction, les notes, etc. seront en français.

Le prix de souscription est fixé à **22.50 florins par volume relié en buckram;** on souscrit à l'ouvrage complet. **Après terminaison le prix sera porté à 100 florins relié en buckram.**

Tiré à 200 exemplaires numérotés à la main.

LIVRES ANCIENS ET MODERNES
EN VENTE AUX PRIX MARQUÉS
CHEZ
MARTINUS NIJHOFF
La Haye, Lange Voorhout 9
adresse télégr.: Books Hague

Prices are in Dutch guilders. One guilder now about $ 0.55

RECENT ACQUISITIONS

Some more extensive divisions: **Archives — Asia — Dairy, Tropical Agriculture, etc. — Memorial publications — Philosophy — Portugal — South and Central America — Surinam**

1. **Alida** et Dorval, ou la nymphe de l'Amstel députée par les Etats-Généraux à la recherche de la liberté. Veropolis (Amst.?) 1785. pet. in-8vo. vieux mar. rouge, dor. s. tr. *Joli ex.*   15.—
   Roman consacré à la recherche de la liberté. Chap. XXVII est intitulé: Dorval va chercher la nymphe en Amérique.
2. **Archives. — Archievenblad, Antwerpsch.** Uitgeg. d. P. Gérard en F. J. van den Branden. Bulletin des Archives d'Anvers. Anvers, 1864–1934. 39 vols. of which 1–13 boards, rest sewed.   120.—
   All published.
3. **— Archievenblad, Nederlandsch.** Groningen, 1892–1937. Year 1–44. 44 vols. In parts.   200.—
   Official organ of the Society of Dutch archivists.
4. **— Archivalia in Italië,** belangrijk voor de geschiedenis van Nederland. Beschreven d. G. Brom. 's-Grav. 1908–15. 3 tom. 4 vol. toile.   15.50
   Cette publication contient de nombr. extraits des pièces décrites. Quelquefois aussi elles ont été réimprimées en entier.
   Rijks Geschiedkundige Publicatiën. Kleine serie, nos. 2, 6, 9. 14.
5. **— Archives** de la Maison-Dieu de Châteaudun. (Publ. p.) A. de Belfort. Paris, 1881. d. vél., n. r.   6.—
6. **— Archives et bibliothèques** de Belgique. Bulletin mensuel de l'Association des conservateurs d'archives, de bibliothèques et de musées. Brux. 1923–38. Année 1—15, livr. 1. En livr.   75.—
   En grande partie épuisé et recherché.
7. **— Archives d'Ypres.** Documents du 16e siècle, faisant suite à l'Inventaire des chartes. Publ. p. I. L. A. Diegerick. (Documents concern. les troubles religieux). Bruges, 1874–77. 4 vol.   10.—
8. **— Archives, Les,** du conseil de Flandre (Raad van Vlaanderen). Publ. par la Ligue nationale pour l'unité belge. Brux. (1926). Av. pl. et facs. 4to.   7.50
   Source importante pour la connaissance du nationalisme flamand publiée par la ligue antiflamande.

ARCHIVES

9. **Archives.** — **Archivo (Real) da Torre do Tombo.** — **Inventario** dos livros das portarias do Reino, 1639–1664. Lisboa, 1909, 12. 2 vol. fol. 25.—
Ces deux volumes renferment l'inventaire des dons faits à des personnes qui pour différentes raisons s'étaient rendues méritoires envers le Portugal. On y rencontre des personnes de différ. nationalités, du Brésil, des Indes, des îles Açores, d'Angleterre, de la Hollande, d'Afrique, etc., etc. Les listes alphabét. des noms et des livres rendent cet Inventaire de grande utilité pour les généalogistes et ceux qui s'occupent de l'histoire de familles.

10. — **Biaudet, H.**, Les archives de Simancas au point de vue de l'histoire des pays du Nord-Baltique. Genève, 1912. W. plan. 4.—
Annales Acad. scient. Fennicae. Ser. B., vol. VIII, nr. 1.

11. — **Devillers, L.**, Inventaire analytique des archives des états de Hainaut. Mons, 1884–1906. 3 vol. gr. in-4to. 30.—

12. — **Diegerick, I. L. A.**, Inventaire analyt. et chronolog. des chartes et documents appartenant aux archives de la ville d'Ypres. Bruges, 1853–68. 7 vol. 12.—

13. — — Inventaire analyt. et chronolog. des chartes et documents, appartenant aux archives de l'ancienne abbaye de Messines (1065—1849). Bruges, 1876. Av. 3 pl. en couleurs de sceaux et de blasons. gr. in-4to. d. veau. 20.—
Fait partie du „Monasticon Flandriae."

14. — **Esnault, G. R.**, Inventaire des minutes anciennes des notaires du Mans (XVIIe et XVIIIe siècles). Publ. et annoté p. L. E. Chambois. Le Mans, 1895–98. 7 tom. 4 vol. d. vél., n. r. 25.—

15. — **Graswinckel, D. P. M.**, Het oud-archief der gemeente Arnhem. 's-Grav. 1935. 3 vol. 15.—
I. Inventaris (–1852). M. voorw. van A. H. Martens van Sevenhoven. — II. Regestenlijst (1113–1543). — III. Brievenlijst (c. 1350–1543), lijst van kaarten en teekeningen, index.
Contient e.a. de nombr. données généalog.
Tirage restreint.

16. — **Guide** internat. des archives. Europe. Paris, 1935. (393 pp.). 12.—
Publication de l'Institut Internat. de coopération intellectuelle.

17. — **Inventaire** des chartes, bulles pontificales, priviléges et documents divers, de la bibliothèque du Séminaire épiscopal de Bruges. Bruges, 1857. 4to. 2.50
Publ. par la Société d'émulation de Bruges.

18. — **Kan, J. van**, Compagniesbescheiden en aanverwante archivalia in Britsch-Indië en op Ceylon. Batavia, 1931. (5.—) 3.—
Cet ouvrage comprend le résultat des recherches, faites dans les archives de Calcutta, Bombay, Madras, Goa, etc. sur des documents (néerlandais et aussi d'autres pays), importants pour l'histoire coloniale néerlandaise.

19. — **Lambin, J. J.**, Tijdrekenkundige lijst van onuitgegeven handvesten, opene brieven enz. berustende onder de archiven der Stadt Ypre, die belangryk zyn voor de plaetselyke geschiedenis. Ypre, 1829. 2.50

20. — **Langlois, Ch. V.**, et **H. Stein**, Les archives de l'histoire de France. Paris, 1891. d. veau. (1000 pp.). *Epuisé.* 15.—
Manuels de bibliographie histor. I.

Prices are in Dutch guilders

20a. **Archives.** — **Meddelanden** från Svenska Riks-Arkivet. Stockholm, 1877–1933. Year 1875–1932 (Series I, 5 vols., 1875–1900; Series II, 7 vols., 1901–1925; Series III, 7 vols.; Series IV, 6 vols., 1926–1932). 25 vols. in 21. Hfmor. 175.—
Complete set of the reports of the Swedish Archives. Scarce.

21. — **Roschach, E.**, Les archives de Toulouse. Histoire du dépôt de l'édifice. Toulouse, 1891. gr. in-4to. 4.—

22. — **Sastachs y Costas, J.**, Memoria s. el Archivo prioral de Cataluña de la Orden de San Juan de Jerusalén. Barcel. 1885. W. plan. 5.—

23. — **Schoonbroodt, J. G.**, Inventaire analyt. et chronolog. des archives de l'abbaye du Val-St-Lambert, Lez-Liége. Liége, 1875, 80. 2 vol. 4to. 10.—

24. — **Schoutheete de Tervarent**, Inventaire général analyt. des archives de la ville et de l'église primaire de St. Nicolas (Waes). Brux. 1872. Av. facs. et 7 pl. de sceaux. 5.—

25. — **Sivré, J. B.**, Inventaris van het Oud-Archief der Gemeente Roermond (–1677). Roermond, 1868–82. 12 parts in 4 vols. Hfcloth. *Very scarce.* 25.—

26. — **Van der Haeghen, V.**, Inventaire des archives de la ville de Gand. Catalogue méthodique général. Gand, 1896. d. rel. 7.50

27. — **Veröffentlichungen** der histor. Kommission der Provinz Westfalen. Inventare der nichtstaatlichen Archive. Münster, 1899–1920. Vol. I–III, 1. Tog. 12 parts. 45.—
I. Kreise Ahaus, Borken, Coesfeld, Steinfurt. — II. Tecklenburg, Warendorf, Ludinghausen, Paderborn. — III. Buren.
Added: Mindener Geschichtsquellen. Bd I. Die Bischofschroniken des Mittelalters. Hrsg. von K. Löffler. — Cosmidromius Gobelini Person, etc. Hrsg. von M. Jansen. — 2 vols.

28. — **Verslagen omtrent 's Rijks oude archieven**, 1878–1937. 's-Grav. 1879–1938. 1st series. 50 vols. W. index. 2d series, vol. 1–10. Tog. 74 vols. sewed (1–6 in 1 vol. boards). 225.—
Annual reports of the State archives of the Netherlands. Complete set, excessively rare.
Added: Verslagen, 1865–1877. 1 vol.

29. **Asia.** — **Brandt, M. von**, Dreiunddreissig Jahre in Ost-Asien. Erinnerungen eines deutschen Diplomaten. Lpz. 1901. 3 vols. W. portr. cloth. 10.—
I. Die preussische Expedition nach Ost-Asien. Japan, China, Siam. Zurück nach Japan. 1862. — II. Japan 1863–1875. In und durch Amerika 1871–1872. — III. China, 1875–1893.

30. — **Bruining, A.**, Bijdr. tot de kennis van de Vedânta. Leiden, 1871. 1.50

31. — **Colleção de monumentos ineditos para a historia das conquistas dos Portuguezes em Africa, Asia, e America** publ. d. Academia Real das Sciencias de Lisboa, sobr. a dir. de R. J. de Lima Felner. la Serie: **Historia da Asia**. Lisboa, 1858–1922. 16 tom. en 11 forts vol. Av. pl. 4to. d. veau. *Très bel ex.* 325.—
I—IV. Lendas da India por Gaspar Correa, cont. as acçoens de Vasco da Gama, F. de Albuquerque, Nuno da Cunha, Duarte de Menezes, F. d'Almeida, e outros, até o ann de 1550. — V. Livro dos pesos, medidas e moedas, p. A. Nunes. — Tombo do estado da India, p. S. Botelho. — Lembranças das cousas da India em 1525. — VI. Decada 13 da historia da India composta p. A. Bocarro, 1612–1617. 2 parties. — VII—IX, XI. Documentos remittidos da India ou livros das Monçoes 1605–1618. 4 vol. — X, XII,

4 ASIA

XIII—XVI. A. de Albuquerque. Cartas seguidas de documentos que as elucidam. 6 vol.
En partie épuisé et rarement complet.

32. **Asia. — Dainelli, G.,** Buddhists and glaciers of Western Tibet. London, 1933. Av. carte et 32 pl. toile. (10.80)   4.50

33. **— Damianus de Goes,** Commentarii rerum gestarum in India citra Gangem a Lusitanis anno 1538. Lovan., R. Rescius, 1539. pet. in-4to. vél.   75.—
Nijhoff–Kronenberg, no. 678. Première édition de cette relation du premier siège de Diu. A la fin un chant à la gloire de D. à Goes par P. Nannius à Alcmar.

34. **— Danvers, E. C.,** The Portuguese in India, being a history of the rise and decline of their empire. London, 1894. 2 vol. Av. de nombr. cartes, pl. et facs. toile orig.   45.—
Epuisé et rare.

35. **— Deussen, P.,** Das System des Vedânta nach den Brahma-Sûtra's des Bâdarâyana und dem Commentare des Çankara über dieselben als ein Compendium der Dogmatik des Brahmanismus. 1883. d. rel.   9.—

36. **— Devéria, G.,** La frontière Sino-Annamite. Description géograph. et ethnograph. d'après des docum. offic. chinois. Trad. pour la 1re fois. Paris, 1886. Av. 15 cartes et de nombr. ill.   18.—

37. **— Etherton, P. T.,** and **H. Hessell Tiltman,** The Pacific, a forecast. London, 1928. W. 10 pl. cloth. (7.50)   2.50

38. **— Etudes asiatiques,** publ. à l'occasion du 25e anniversaire de l'Ecole française d'Extrême-Orient. Paris, 1925. 2 vol. Av. 59 pl. et de nombr. ill. dans le texte. (25.—)   16.—
Contient e.a.: **P. Pelliot,** Quelques textes chinois concern. l'Indochine hindouisée. — **J. Ph. Vogel,** Gangā et Yamunā dans l'iconographie brahmanique. — **P. Demiéville,** La musique čame au Japon. — **H. Maspéro,** Le roman de Sou Ts'in. — **G. Soulié de Morant,** Le problème des bronzes antiques de la Chine. — **A. Foucher,** Notes sur l'itinéraire de Hiuan-Tsang en Afghanistan. — etc.

39. **— Featherstone, B. K.,** An unexplored pass. Narrative of a thousand-mile journey to the Kara-koram Himalayas. London, 1926. Av. carte et 25 reprod. toile. (10.80)   4.—

40. **— Formation** of the Singapore institution, 1823. Malacca, Printed at the Mission Press, 1823. 8vo. d. veau fauve. (110 pp.). *Très rare.*   48.—
Sir Th. S. Raffles fut un des principaux fondateurs de l'Institut de Singapore, qui avait pour but „the cultivation of the languages of China, Siam and the Malayan Archipelago, and the improvement of the moral and intellectual condition of the inhabitants of those countries". Dans l'écrit nommé ci-dessus on trouve e.a.: Minute by Raffles on the establishment of a Malayan college at Singapore. — The Anglo-Chinese college for the cultivation of Chinese and English literature, etc. founded at Malacca, 1818. — Suggestions by Dr. Morrison respect. the Institution at Singapore. — etc.

41. **— Ghistele, Joos van,** Tvoyage.... Tracterende van veelderande wonderlicke ende vremde dijnghen, gheobserueerd over d'zee, in den landen van Sclavonien, Griecken, Turckien, Candien, Rhodes ende Cypers. Voords ooc in den lande van Beloften, Assirien, Arabien, Egypten, Ethyopien, Barbarien, Indien, Perssen, etc. Ghendt, H. van den Keere, 1557. 4to. d. veau. (Rel. mod.).   175.—
Bibl. Belgica, G 75. „Le voyage de Josse van Ghistele est le plus remar-

Prices are in Dutch guilders

ASIA 5

quable et le plus important de tous les voyages en Orient faits au moyen-
âge. Aucun de ses devanciers ne donne des détails aussi exacts et aussi
circonstanciés sur les contrées visitées.... Malgré la supériorité de
l'ouvrage de van Ghistele sur tous ceux du même genre, il n'a été reproduit
dans aucune des grandes collections de voyages."
Edition originale.
Ex. en bon état; seulement une tache dans quelques marges.

43. **Asia. — Jourdain Catalani de Sévérac** (14e siècle), Mirabilia de-
scripta. Les merveilles de l'Asie. Texte latin, facsimile et trad.
francaise av. introd. et notes par H. Cordier. Paris, 1925. Av.
fasc. des 19 pl. du ms. latin. 4to.   10.—

44. — **Jung, E.**, Questions d'Orient. L'Islam sous le joug. (La nouvelle croisade). Paris, 1926.   1.—

45. — — L'Islam et l'Asie devant l'Impérialisme. Paris, 1927.   1.50

46. — **Menant, D.**, Les Parsis. Histoire des communautés zoroastrien-
nes de l'Inde. Paris, 1898. 1re partie (seule parue). Av. 21 portr.
et ill. *Epuisé.*   20.—
Annales du Musée Guimet. Bibliothèque d'études. T. 7.

47. — **Mission Pavie Indo-Chine,** 1879–1895. Paris, 1898–1919. 10 vol.
Av. 69 cartes, une foule de pl. en couleurs et en noir et ill. et
Atlas de 10 cartes en couleurs. Ens. 11 vol. 4to. d. chagr. (2 vol. br.).
                                                                     175.—
A. Géographie et voyages. I, II. Exposé des travaux de la mission. Par
A. Pavie. 2 vol. — III, IV. Voyages au Laos, au centre de l'Annam et
chez les sauvages du Sud-Est et de l'Est de l'Indo-Chine. Par Cupet, de
Malglaive et Rivière. Introd. par A. Pavie. 2 vol. — V. Voyages dans le
haut Laos et sur les frontières de Chine et de la Birmanie. Par Lefèvre-
Pontalis. Introd. par A. Pavie. — VI. Passage du Mé-Khong au Tonkin.
Par A. Pavie. — VII. Journal de marche (1888–89). Evénements du Siam
(1891-93). Par A. Pavie. — *B.* Etudes diverses. I. Recherches sur la littérature du Cambodge, du Laos et du Siam. Par A. Pavie. — II. Recherches
sur l'histoire du Cambodge, du Laos et du Siam. Par A. Pavie. — III.
Recherches sur l'histoire naturelle de l'Indo-Chine Orientale. Par A. Pavie.
Ex. ouvrage de la plus haute importance et épuisé.

48. — **Monfried, H. de,** Pearls, arms and hashish. Pages from the
life of a Red Sea navigator. Coll. by I. Treat. London, 1930.
W. 16 pl. cloth. (10.80)   3.50

49. — **Morden, W. J.**, Across Asia's snows and deserts. London, 1927.
W. 65 pl. cloth. (12.60)   4.50
Narrative of the Morden-Clark expedition across Asia from Bombay
to Peking.

50. — **Radde, G.**, Reisen an der Persisch-russischen Grenze. Talysch
und seine Bewohner. Lpz. 1886. Av. carte, 4 pl. et plus. ill. toile.
                                                                      4.—

51. — **Riebeck, E.,** Die Hügelstämme von Chittagong. Ergebnisse
einer Reise im J. 1882. Berlin, 1885. Av. carte et 21 pl., dont 2
en couleurs de tissus. fol. cart.   18.—

52. — **Scheref,** Scheref-Namch ou histoire des Kourdes. Texte persan
publ., traduit et annoté par V. Veliaminof-Zernof. St. Pétersbourg, 1860, 62. 2 vol.   15.—
Marque de bibliothèque sur les titres.

53. — **Schillinger, F. C.**, Persianische und Ost-Indianische Reis,
1699–1702. Nürnberg, 1707. W. front., 2 maps and 8 pl. Hfvellum.   100.—
Original edition. The author visited Persia, India (Suratte) and Ceylon.

Mart. Nijhoff, The Hague — Cat. No. 633

54. **Asia. — Strong, A. L.,** Red star in Samarkand. N.-York, 1929.
W. 9 pl. cloth. (8.75)   3.—
   Contents: The iron road to Turkestan. — The cotton empire. — The
   profane invasion of Holy Bokhara. — Making Bolsheviks. —· Martyrs
   for women's rights. — etc.

55. **— Thalasso, A.,** Anthologie de l'amour Asiatique. Paris, 1907.
                                                          2.50

56. **— Tod, J.,** Annals and antiquities of Rajast'han, or, the Central
and Western Rajpoot States of India. London, 1914. 2 vols. W.
maps. cloth.   6.—

57. **— Weil, G.,** Geschichte der Chalifen. Nach handschriftl. grössten-
theils noch unbenützten Quellen bearb. Mannheim, 1846–51. 3 vol.
cart.   10.—

58. **— Witsen, N.,** Noord- en Oost-Tartarye, ofte bondig ontwerp van
eenige dier landen en volken, welke voormaels bekent zijn geweest.
Beneffens verscheide.... meest nooit voorheen beschreve Tar-
tersche en nabuurige gewesten.... in de Noorder en Oostelyk-
sche gedeelten van Asia en Europa. 2e dr. Amst., F. Halma, 1705.
2 tom. 1 vol. Av. 2 front., 13 cartes, 40 pl. et plusieurs grav. en
taille-douce dans le texte. fol. d. veau.   175.—
   Ouvrage fort intéressant dans lequel on trouve des notices sur les langues
   et les coutumes des peuples qui habitent le Nord, le Nord-Est et le Nord-
   Ouest de l'Asie, la Nouvelle Zélande, la Nouvelle-Guinée, l'Amérique, etc.
   Edition très rare; la première n'a jamais été mise dans le commerce et est
   introuvable.
   Les pp. 19/20 de la préface, les pp. 209/210 du texte et la table à la fin
   sont en ms. très lisible.

59. **Blosius, Lud.,** (Dacryanus), Speculum monachorum a· Dacryano
ord. s. Benedicti abbate conscriptum. Antehac nusquam excusam.
Venundantur a Bartholomaeo Gravio Louanij, 1538. (*À la fin:*)
Typis Seruatii Zaseni Diestensis, 1538. 4to.   25.—
   Nijhoff–Kronenberg, no. 427. Voir aussi Biographie nationale de Bel-
   gique. II (1868), 505.

60. **Bruining, H.,** en **P. Hofstede,** Het Kralinger lasterschrift, gen.
Klaare en grondige wederlegging enz. een uitvoerige beantwoor-
dinge onwaardig. Rott., J. Bosch, 1757. veau écaillé, bordure sur
les plats, dos et tr. dor. *Ex. sur papier fort.*   2.50
   Contestation religieuse à Rotterdam.

61. **Chronology. — Calendrier de Cordoue, Le,** de l'année 961. Texte
arabe et ancienne trad. latine publ. p. R. Dozy. Leyde, 1873.
Epuisé et rare.   7.50

62. **— Roberts, R.,** Onder verbeteringe. Korte inleydinge der feesten
'Israels, twelck rechte tijtkaarten zijn, waer in ghy sien meucht
hoe veel groot jaren die werelt ghestaan heeft .... noch staan sal,
ende in wat groot jaar datse vergaan sal. Amst., R. Roberts,
1593. W. 14 maps and coloured chronolog. figg. 4to. vellum.
                                                         30.—
   This work contains an explication of some passages of the Old Testament
   and some Israelitic festivities with regard to chronology. At the end some
   poems.
   See Moes en Burger. Vol. III, p. 61.

63. **— Wijk, W. E. van,** Le nombre d'or. Etude de chronologie technique. Suivie du texte de la Massa compoti d'Alexandre de Ville-

dieu avec traduction et commentaire. La Haye, 1936. Av. 17
pl. et figg. 10.—
La première partie du livre traite de l'origine du comput pascal, que
l'auteur envisage sous des points de vue entièrement originaux; elle sert
d'introduction à la deuxième, qui contient le texte d'un comput versifié,
la **Massa Compoti** d'Alexandre de Villedieu, texte très répandu dans les
manuscrits, et qui pourtant n'a jamais été imprimé.
Le texte est·accompagné — en plus d'une traduction fidèle — d'un commentaire qui, suivi d'un index alphabét., rend le livre vraiment utile comme
manuel de chronologie technique médiévale.

64. **Chronology. — Wĳk, W. E. van**, Decimal tables for the reduction of
Hindu dates from the data of the Sūrya-Siddhānta. The Hague,
1938. W. portrait and tables. 4.—
Hindu chronology always remains an intricate matter to the European
mind. To enable philologists to effectuate themselves the calculations
necessary for the reduction of a given *tithi* to its Julian equivalent, the
above mentioned tables have been dressed. The little book will however
be also found of interest to students of chronology as it forms a concise
introduction to the most complicate form of calendarical reckoning, the
one based on the true movements of heavenly bodies.

65. — — De gregoriaansche kalender. Een technisch-tijdrekenkundige studie. 's-Grav. 1938. W. 12 portr. and pl. 4.—

66. **Cicero**, Pleitrede voor Publius Sulla. Vert. d. J. Cochez. Leuven,
1929. 1.—
Philologische studien. Teksten, no. 1.

67. **Crimes. — Baelde, R.**, Studien over godsdienstdelicten. 's-Grav.
1935. toile. 6.50
Drie Attische godsdienstprocessen. — Het godslasteringsdelict in de
wetgeving en in de juridische theorie tot aan de Aufklärung. — De be-
strijding van godslasterlijke uitingen in ons land tot het einde der 18e
eeuw. — De meeningsstrijd omtrent de godslastering tijdens de Aufklärung. — De godslasteringsbepalingen in ons Wetboek van Strafrecht.
Strafrechtel. en criminolog. onderzoekingen. V.

68. — **Drukker, L.**, De sexueele criminaliteit in Nederland, 1911–1930.
's-Grav. 1937. W. 2 maps and tables. cloth. 5.—
Strafrechtel. en criminolog. onderzoekingen. VIII.

69. — **Larson, J. A.**, a.o., Lying and its detection. With introd. by
A. Vollmer. Chicago, 1932. Av. pl. et tabl. toile. (16.50) 7.50
Ancient and modern forensic methods for the detection of the innocence
or guilt of the suspect. — Cardio-pneumo-psychograph. experiments, etc.
Pp. 419–442: Bibliography.

70. — **MacDonald, J. C. R.**, „Crime is a business". Buncos, rackets,
confidence schemes. Stanford University, 1939. W. pl. cloth. 6.—
„This book is the most complete exposition of the methods used by
criminals for defrauding the public that has ever been presented. It has
been written by an experienced police officer."

71. — **Misdadigheid** en wangedrag in verband met het verschijnsel
zwakzinnigheid. Lezingen (d. D. Wiersma, E. A. D. E. Carp,
J. M. van Bemmelen e.a.). 's-Grav. 1939. cloth. 4.20
Strafrechtel. en criminolog. onderzoekingen. X.

72. — [**Pièces, Réimpressions de**, des 16e et 17e siècles, racontant des
crimes, des miracles, etc. S.-Germain-en-Laye, etc.]. 1874, 75. Ens.
9 pièces en 1 vol. pet. in-8vo. mar. bleu, dos orné, fil., dent. intér.,
tête dor., n. r. 40.—
Discours estrange et pitoyable d'une femme envers ses enfans a l'accasion
(*sic*) d'un faux monnoyeur et pour la nécessité d'elle et de ses dits enfants,
laquelle s'est desespérée et pendue, etc. Le tout véritable est approuvé

et advenu, auprès de Rouen en un village nommé La Ferté en Bray. Paris, 1608. — **La grande cruauté** de massacre arrivé depuis n'aguères en la ville du Mans par une femme qui a esgorgé deux de ses filles, laquelle a esté bruslée en la Place au Laict, le 15e Oct. 1609. Lyon, 1610. — **Histoire prodigieuse et pitoyable** d'une jeune homme qui a tué et bruslé sa propre mère au village de Nogent sur Marne près Paris. Avec la punition qui en a esté faicte. Paris, 1611. — **La conversion** de dix notables personnes a la foy et religion catholique, apostolique, et romaine en la ville de Grasse en Provence, confirmée par un évident miracle. Paris, 1612. — **Exécution** d'un capitaine dans la ville de Lyon. Ens. la desloyauté d'une damoiselle envers son mary. Paris, 1626. — **La conversion publique** de quatre personnes de qualité faicte en l'église Sainct André des Arts en présence de plus de quatre mille assistans le Dimanche 17 Novembre. Paris, 1619. — **Miracle** advenu en la ville de Lyon en la personne d'un jeune enfant, lequel ayant esté mort vingt-quatre heures est ressuscité par l'intercession de la Sacrée Vierge. Lyon, 1619. — **Histoire espouvantable et veritable** arrivée en la ville de Soliers en Provence d'un homme qui s'estoit voué pour estre d'esglise et qui n'ayant accomply son voeu le diable lui a couppé les parties honteuses et couppé encore la gorge à une petite fille aagée (*sic*) de deux ans ou environs. Paris, 1619. — **Discours véritable** d'un usurier lequel miraculeusement a esté mangé des rats a Charret proche la ville d'Aix en Provence le 2e Aoust 1606. Lyon, 1606.
Exx. sur peau de vélin.

73. **Crimes.** — **Pitaval, G. de,** Causes célèbres et intéressantes avec les jugemens qui les ont décidées. La Haye, J. Neaulme, Paris, 1737–45. 22 vols. W. front. sm. 8vo. old calf.   45.—

74. — **Röling, B. V. A.,** De wetgeving tegen de z.g. beroeps- en gewoontemisdadigers. 's-Grav. 1933. toile. (12.—)   6.—
L'ouvrage lui-même est suivi d'une partie supplément., comprenant non moins de 230 pp., qui contient, de 85 pays, le texte des lois contre les criminels de profession et de coutume, qui y sont en vigueur, toutes en langue originale. Nulle part on ne trouve réuni les textes de ces lois. Strafrechtel. en criminolog. onderzoekingen. III.

75. **Dairy, Pisciculture, Sericulture, Sheep-breeding, Tropical agriculture, Viniculture.** — **Annales de la Société séricicole,** pour l'amélioration et la propagation de l'industrie de la soie en France. Paris, 1837–49. T. I–XII. Av. table générale des tom. I–X. Ens. 10 vol. Av. pl. d. veau.   55.—

75a.— **Bassi, A.,** Del mal del segno calcinaccio o moscardino malattia che affligge i bachi da seta e sul modo di liberbarne le bigattaje. Lodi, 1835, 36. 2 tom. 1 vol. cart., n. r. *Bel ex.*   4.50

76. — **Congrès internat.** d'aquiculture et de pêche. Paris 1900. Mémoires et comptes-rendus des séances. Paris, 1901. W. pl.   7.50

77. — — 7e, d'aquiculture et de pêche. Paris 1931. Rapports, etc. Orléans, 1931. 2 vol.   8.50

78. — — 3e, de défense contre la grêle et congrès de l'hybridation de la vigne. Lyon 1901. Compte rendu. Lyon, 1992. 2 vols. Av. carte.   12.50

79. — — 3e, de laiterie. La Haye, Schéveningue 1907. Compte-rendu des travaux et des excursions. Amst. 1907. Av. pl. et ill.   5.—

80. — — 1er, du maïs. Pau 1930. Comptes rendus. Pau, 1933. Av. ill.   4.50

81. — **Congrès** du mouton. Paris 1929. Rapports et discussions. Monographies des races ovines. Paris, 1930. 2 vols.   6.—

Prices are in Dutch guilders

DAIRY 9

82. **Dairy, Pisciculture, Sericulture, Sheep-breeding, Tropical agriculture, Viniculture.** — **Congrès internat.**, 6e, d'oléiculture. Nice 1923. Compte rendu des travaux. (Les ravageurs de l'olivier. Insectes et cryptogames. Les traitements pratiques). Paris, 1924. 6.—
83. — — 1er, du raisin et du jus du raisin. Tunis 1936. Procès-verbaux. Tunis, 1937. Av. portraits et pl. 7.50
84. — — **séricicole.** Paris 1878. Comptes rendus sténograph. Paris, 1879. 4.—
85. — **Congres,** 9e, van het Algemeen Syndicaat van suikerfabrikanten in Nederl.-Indië. 1911. Handelingen. Soerabaia, 1911. 2 tom. 1 vol. Av. pl. cart. 18.—
86. — — 7e, van het Algemeen Syndicaat van suikerindustrie op Java. Soerabaia 1905. Handelingen. Soerabaia, 1905. Av. pl. cart. 12.—
87. — — **Thee-,** met tentoonstelling. Bandoeng 1924. Handelingen. Weltevreden, s.d. Av. 47 pl. toile. (12.—) 4.50
88. — **Congreso internac. filoxérico.** Zaragoza 1880. Zaragoza, 1880. veau. 10.—
89. — — 7°, de oleicultura. Sevilla 1924. Madrid 1926. W. numer. portr. and ill. 4to. 14.—
90. — **Dufour, Ph. S.,** Traitez nouveaux et curieux du café, du thé et du chocolate. Lyon, 1685. Av. front., 3 vign. et 3 pl. pet. in-8vo. veau. 15.—
91. — **Francus, J.,** Veronica thee'zans i. e. collatio veronicae Europaeae cum theé chinitico. Acc. mantissae loco conjectura de Alysso Dioscoridis. Lps. (v. 1700). Av. front. et 3 pl. de drogues. 12mo. d. vél. 15.—
92. — **Freschi, G.,** Guida per allevare i bachi da seta. 2a ed. accresc. San-Vito, 1840. Av. 1 fig. cart. 2.50
93. — **Jotemps, Perault de, Fabry** und **Girod,** Ueber Wolle und Schaafzucht. A. d. Franz. und n. d. gegenwärt. Standpuncte der Woll-und Schaafkenntniss in Deutschland bearb. von A. Thaer. Berlin, 1825. d. veau. 3.—
94. — **Kongress, 11er Milchwirtschaftlicher Welt-.** Berlin 1937. Berichte. Hildesheim, 1938. 4 vol. Av. ill. toile. *Epuisé et rare.* 50.—
95. — **Neszler, L.,** Die Bereitung, Pflege und Untersuchung des Weines besond. für Winzer, Weinhändler und Wirte. 6e verm. Aufl. Stuttgart, 1894. Av. ill. 1.75
96. — **Petri, B.,** Das ganze der Schafzucht in Hinsicht auf unser deutschen Klima und der angrenz. Länder, insbes. von der Pflege, Wartung und den Eigenschaften der Merinos und ihrer Wolle. Wien, 1815. Av. 16 pl. 7.50
97. — **Reyntkens, J. B.,** Den sorghvuldighen hovenier ende de oprechte pracktycke om blommen te zaeyen, planten, ende gouverneren naer de conste ende daghelijcksche hofbauwinghe. Verm. met den Nederlantschen hesperides, d.i. oeffeningh en ghebruyck vande l i m o e n e n o r a n i e b o o m e n. 2e dr. Ghendt, 1695. sm. 8vo. hfvellum. (Mod. binding). 24.—
98. — **Stella, B.,** Il tabacco. . . . dell' origine, historia, coltura, preparatione, qualita, e uso in fumo, in polvere, in foglia, in lambi-

Mart. Nijhoff, The Hague — Cat. No. 633

tivo, et in medicina della pianta volgarmente detta tabacco. Roma, 1669. W. 3 woodcuts. sm. 8vo. vellum. *Scarce.*   30.—
One leaf slightly damaged.

99. **Dairy, Pisciculture, Sericulture, Sheep-breeding, Tropical agriculture, Viniculture.** — **Stevens, K.**, ende **J. Liebaut,** De veltbouw ofte lantwinninghe inhoudende ....cruydthoven ende fruythoven te maecken,.... byen te houden, te distilleren,.... visschen te vanghen,.... wijngaerden te oeffenen, medicinale wijnen te bereyden, parck voor wilde beesten te maecken, midtsg. de wolve jacht. Verm. d. M. Sebizius Silesius. Amst., C. Claesz, 1594. W. large coloured engraving on titlepage and woodcuts in the text.fol. limp vellum.   90.—
Moes en Burger, II, p. 100, no. 351. A very fine and clean copy. In this state very scarce.

100. — **Thaer, A.**, Handbuch für die feinwollige Schaafzucht. Berlin, 1811.   3.50

101. — **Tonduz, A.**, La enfermedad del cafeto. — **Id.**, La enfermedad cacaotero. — **Id.**, La fumagino del cafeto. — San José, 1893–97. 3 pieces. *Reprint.*   2.50

102. — **Trémols y Borrell, F.**, Informe acerca de las cepas de los Estados-Unidos de America ....los recursos que pueden prestarnos para la repoblacion de los viñedos destruidos por la filoxera. Barcelona, 1881.   5.—

103. — **Ukers, W. H.**, All about coffee. 2d ed. N. York, 1935. Av. front. en couleurs et de nombr. ill. toile.   30.—
Historical — Technical — Scientific — Commercial — Social — Artistic. „Here for the first time have been assembled, in their right order, all the essential facts about coffee." Les pp. 733 à 788 contiennent: A coffee chronology; a dictionary of coffee; a coffee bibliography.

104. — **Voorschriften, Nog nooit geopenbaarde,** voor tabaksfabrikeurs, tabaksrookers om inlandsche rook- en snuiftabak te vervaardigen. N. h. Duitsch. 2e dr. Zutphen, 1805. pet. in-8vo.   5.—
Schotel, no. 124.

105. — **Wagner, J. Ph.**, Beitr. z. Kenntniss und Behandlung der Wolle und Schaafe. Nebst Verzeichniss mehrerer Schäfereien. Berlin, 1820. cart.   4.50

106. **Does, J. van der,** Geluck wenschinge van de Haegsche schutterye aen den Prince van Orange, wederkeerende in 's Gravenhage, na het veroveren van verscheyde steden en sterckten. 's Grav., D. Geselle, 1673. W. woodcut on titlepage. 4to.   5.—
Knuttel, no. 10761.

107. **Einsiedeln.** — **Hartmannus, Chr.**, Annales Heremi Deiparas Matris monasterii in Helvetia ordinis S. Benedicti, antiquitate, religione, frequentia, miraculis, toto orbe celeberrimi. Friburg, 1612. W. engraved titlepage, 1 pl. and numer. finely engraved coats of arms in the text. fol. calf, with coats of arms on both sides. (Corners of the binding somewhat damaged).   65.—
Standardwork on the monastery of Einsiedeln.

108. **Ernst, V.**, Die Entstehung des niederen Adels. Stuttgart, 1916.   1.50

109. **Ex-libris.** — **Archives** de la Société (française) des collectionneurs d'ex-libris. Paris, 1893–1937. Année 1–44. Av. table des années 1–20. 42 vol. Av. de très nombr. pl. et ill. pet.-in-fol. En livr. 375.—
Revue de la plus grande importance, constituant un des plus beaux répertoires d'ex-libris et formant en même temps une mine extrêmement riche de documents héraldiques, généalogiques et historiques. Il n'y manque qu'une pl. Extrêmement rare en état complet.

110. — **Vorsterman van Oijen, A. A.,** Les dessinateurs néerlandais d'ex-libris. Arnhem, 1910. Av. 70 pl., conten. 342 reprod. d'ex-libris. fol. toile. *Epuisé.* 20.—

111. **Freemasonry.** — **Krause, K. C. F.,** Die drei ältesten Kunsturkunden der Freimaurerbrüderschaft. Dresden, 1820, 21. 2 vol. d. veau. *Très rare.* 36.—

112. **Geschichte** der Wissenschaften in Deutschland. Neuere Zeit. München, 1864–1912. 24 tom. 30 vol. d. veau fauve unif. *Bel ex.* 450.—
1. **Bluntschli,** Geschichte der neueren Staatswissenschaft. 1864. — 2. **v. Kobell,** Gesch. der Mineralogie. 1864. — 3. **Fraas,** Gesch. d. Landbauund Forstwissenschaft. 1866. — 4. **Peschel,** Gesch. d. Erdkunde. 1865. — 5. **Dorner,** Gesch. d. protestant. Theologie. 1867. — 6. **Werner,** Gesch. d. Kathol. Theologie. 1866. — 7. **Lotze,** Gesch. d. Aesthetik. 1867. — 8. **Benfey,** Gesch. d. Sprachwissenschaft und oriental. Philologie. 1869. — 9. **v. Raumer,** Gesch. d. german. Philologie. 1870. — 10. **Kopp,** Die Entwicklung der Chemie in den neueren Zeit. 1873. — 11. **Karmarsch,** Gesch. d. Technologie. 1872. — 12. **Carus,** Gesch. d. Zoologie. 1872. — 13. **Zeller,** Gesch. d. Philosophie. 1873. — 14. **Roscher,** Gesch. d. Nationalökonomie. — 15. **Sachs,** Gesch. d. Botanik. 1876. — 16. **Wolf,** Gesch. d. Astronomie. 1877. — 17. **Gerhardt,** Gesch. d. Mathematik. 1878. — 18. **Stintzing,** Gesch. d. deutschen Rechtswissenschaft, 1880–1910. 3 tom. 5 vol. — 19. **Bursian,** Gesch. d. klassischen Philologie. 1884. — 20. **v. Wegele,** Gesch. d. Historiographie. 1885. — 21. **Jähns,** Gesch. d. Kriegswissenschaften. 1889–91. 3 vol. — 22. **Hirsch,** Gesch. d. medizin. Wissenschaften. 1893. — 23. **Zittel,** Gesch. d. Geologie u. Paläontologie. 1899. — 24. **Gerland,** Gesch. d. Physik. 1913.
Tous les volumes en impression originale.

113. **Heretics.** — **Castellion, S.,** Traité des hérétiques à savoir, si on les doit persécuter et comment on se doit conduire avec eux, selon l'avis, opinion, et sentence de plusieurs auteurs. Ed. nouv. par A. Olivet. Genève, 1913. 2.—

114. **Herring-fishery.** — **Gevers Deynoot, W. Th.,** De magno s. halecum piscatu belgico (haringvisscherij). L. B. 1829. cart. 2.50

115. **Hockett, H. C.,** Political and social history of the United States, 1492–1828. N. York, 1925. W. 36 maps. cloth. (7.50) 3.—

116. **Hooft, P. Czn. — Haan, J. C. de,** Studien over de Romeinsche elementen in Hooft's niet-dramatische poëzie. Antw. 1923. 1.50

117. **Horatius. — Burger Jr., C. P.,** „Aere perennius". Scherts en ernst in de oden van Horatius. 's-Grav. 1926. toile. (4.—) 2.—

118. — **Olivier, F.,** Les épodes d'Horace. Lausanne, 1917. 3.—

119. **Intrigues. — Wassenaer JCz., G. van,** Heerschappye en cuypconsten om tot staet en bedieninge te comen. Bedekte konsten in regeringen en heerschappien.... waer door koningen en princen, edelen en steden ...., haer ....heerschappie vast stellen. M. konsten om staet en bedieningen te bekomen in regeringen en hoven van princen en heerschappien. Utrecht, G. van Zyll en D.

Mart. Nijhoff, The Hague — Cat. No. 633

van Ackersdyck, 1657. W. nice front. after C. Ulft by C. van Dalen. 12mo. old vellum.   30.—
Very scarce booklet, dealing with the intrigues, carried on by princes, noblemen, etc. in order to consolidate their power, as well as by the lower classes in order to work themselves up.

120. **Jaarverslag der Rijkscommissie tot het opmaken en uitgeven van een inventaris en eene beschrijving van de Nederl. monumenten van geschiedenis en kunst,** over 1903–1918. ('s-Grav. 1904-18). 15 vols.   40.—
Rapports annuels de la Commission de l'Etat pour la publication d'un inventaire et d'une description des monuments néerlandais d'histoire et des beaux arts.
Collection complète.
Continué par: **Jaarverslag** der Rijkscommissie voor de monumentenzorg over 1918 en 1919. ('s Grav. 1920). 2 vol. Av. 34 pl. 4to.
Tout ce qui a paru.

121. **Kolff, B.,** Autonoom havenbestuur. Onderzoek naar het bestuur der haven van Rotterdam aan de hand van Britsche en Fransche gegevens. Leiden, 1928. (2.90)   1.50

122. **Labé, Louize,** Sonnetten. Naast den oorspronk. tekst vert. d. P. C. Boutens. Maastricht, 1924. W. portr. In a full red mor. binding.   45.—
Beautiful copy of this edition limited to 215 copies, of which only 175 in the trade.
Trajectum ad Mosam. Vol. 7.

123. **(La Roche-Guilhem, Mlle De),** Histoire des favorites, ...sous plusieurs regnes. Par Mlle D***. Amst., P. Marret, 1700. 2 tom. 1 vol. Av. front. et 10 portr. pet. in-8vo. veau, dos dor.   10.—
Marie de Padille sous Pierre le Cruel, roi de Castille. — Agnes Soreau sous Charles VII, roi de France. — Roxelane sous Soliman II, empereur des Turcs. — Marozie sous plusieurs Papes, etc.

124. **Le Long, J.,** Kort historisch verhaal van den eersten oorsprong der Nederl. Gereformeerden kerken Onder 't kruys; beneff. alle leer- en dienstboeken soo van de Nederduytsche, als Fransche gemeentens, en ders. veranderingen tot na de Reformatie. M. oorspr. stukken. Amst., Wed. S. Schouten en Zn., 1751. 4to. vél.   20.—
Ouvrage fort rare et très intéressant, formant une histoire de l'église réformée des Pays-Bas dite *sous la croix*, et notamment des communes établies par les réfugiés français, belges et hollandais à Londres et à Francfort, accomp. d'une histoire littéraire et d'une bibliographie de ses livres liturgiques.

125. **Leopold, J. H.,** Verzamelde verzen. Uitgeg. d. P. N. van Eyck. Rott. 1935. W. portrait. roy. 8vo. vellum.   25.—
One of the 50 copies on „Ossekop-van Gelder", of which it is no. 16.

126. **Lof, Veeler wonderens wonderbaarelijck,** behels. het lof van het podagra, het zwemmen, den oliphant, etc. Amst. 1664. 3 tom. 1 vol. Av. 2 front. et 8 pl. 12mo. vél.   15.—
Contient: Het hatelick podagra; 't lof des vloos; het lijf-bergende zwemmen; dat gruwelick groot beest den oliphant; die mensch-lievende luys; dat hert-kittelende lachen; die plaeghlicke derden-daeghse-koorts; dien alder-treffelicksten uyl; etc.

127. **Looff, J. — Man, M. G. A. de,** Het leven en de werken van Joh. Looff, stempelsnijder en graveur te Middelburg. Middelburg, 1925. Av. 31 ill. sur 10 pl. *Extr.*   2.—

Prices are in Dutch guilders

128. **Lijst, Voorloopige,** der Nederlandsche monumenten van geschiedenis en kunst. Utrecht, 1908–33. 11 vols. in 13. 45.—
Provisional (official) list of Netherlands monuments of history and art (per provinces): I. Utrecht. — II. Drente. — III. Zuid-Holland. — IV. Gelderland. — V*a*. Noord-Holland. — V*b*. Amsterdam. — VI. Zeeland. — VII. Overijssel. — VIII. Limburg. 2 vol. — IX. Friesland. — X. Noord-Braband. — XI. Groningen.
Complete collection. Vol. 1—4 are out of print.

129. **Manners and customs.** — **Bing, V.,** en **Braet von Ueberfeldt,** Nederlandsche zeden en gebruiken. Amst. 1857–59. 6 parts, contain. 6 ll. of text and 18 coloured lithographs. large fol. Original covers preserved. 90.—
De volkstrekschuit. — Eene Hollandsche boerenbruiloft. — De beugelbaan. — De zoare paol (Drenthe). — De kermis te Egmond aan Zee. — De kaasmarkt te Alkmaar. — De vischafslag aan het strand. — Het vogelschieten op Walcheren. — Het ringrijden (Zeeland). — Eene kaasmakerij in Noord-Holland. — Melkschuit. — etc.
One of the very rare copies with superior colouring.

130. — **Champfleury,** Histoire de l'imagerie populaire. Paris, 1869. Av. 38 ill. d. rel. Bradel, n.r., couv. cons. 3.50
Le Juif-Errant. — Histoire du bonhomme Misère. — La Danse des morts en 1849, etc.

131. — **Folklore brabançon, Le.** Brux. 1921–38. Année 1–17 (= nos. 1–100). 17 vol. Av. pl., ill. et figg. En livr. 150.—
Très rare. Les titres n'ont jamais paru.

132. — **Franklin, A.,** La civilité, l'étiquette, la mode, le bon ton du 13e au 19e siècle. Paris, 1908. 2 vol. d. mar. vert. 10.—

133. — — La vie privée au temps des premiers Capétiens. Paris, 1911. 2 vol. d. mar. vert. 8.—

134. — **Fuchs, E.,** Illustrierte Sittengeschichte vom Mittelalter bis zur Gegenwart. München, (v. 1900). Mit 3 Ergänzungsbänden. Ens. 6 vol. Av. de nombr. pl. en couleurs et en noir et des ill. 4to. veau orig. 70.—
I. Renaissance. — II. Die galante Zeit. — III. Das bürgerliche Zeitalter.

135. — **Ostwald, H.,** Die Berlinerin. Kultur- und Sittengeschichte Berlins. Berlin, 1921. Av. 343 ill., pour la plupart d'après d'anciennes grav. toile. 4.50
Die Damen. — Dienstboten. — Berliner Kinder. — Kleinbürgertum und Proletariat. — Halbwelt. — etc.

136. — **Revue d'ethnographie.** Publ. sour la dir. de Hamy. Paris, 1882–89. 8 vol. Av. pl. et ill. d. mar. rouge. 85.—
Tout ce qui a paru.

137. — **Revue d'ethnographie** et des traditions populaires. Paris, 1920–29. 10 années (= 40 nos.). 10 vol. Av. pl. En livr. 75.—
Tout ce qui a paru. Les t. 7-10 sans titres et tables (très probablement pas parus).

138. — **Revue des traditions populaires.** Recueil mensuel de mythologie, littérature orale, ethnographie traditionelle et art populaire. Paris, 1886–1919. Année 1—34. Av. table des années 1886–1893. Ens. 35 tom. 32 vol. Av. ill. dont 18 vol. d. veau, le reste en livr. 375.—
Tout ce qui a paru. Le titre et la table de l'année 1919 n'ont pas paru.

139. — **Schotel, G. D. J.,** Het maatschappelijk leven onzer vaderen in de 17e eeuw. Haarlem, 1868. 2 vol. Av. 3 pl. gr. in-8vo. 3.—

Mart. Nijhoff, The Hague — Cat. No. 633.

140. **Manners and customs. — Streso, J. A.**, en **Jac. van Manen**, Levens-
wijze en gewoonten onzer voorvaderen tot het einde der 16e
eeuw. Haarlem, 1814. 4to. cart.   3.50

141. — **Taxandria.** Tijdschrift voor Noord-Brabantsche geschiedenis
en volkskunde. Bergen-op-Zoom, 1894–1933. Year 1—40. 40 vols.
W. pl. of which 24 vols hfcloth, the rest cloth.   225.—

142. — **Vlaamsch leven.** Zelfstandig Vlaamsch geïllustr. weekblad.
Brussel, 3 Oct. 1915 — 29 Sept. 1918. Année 1—3. 3 vol. Av. de
nombr. portr. et ill. gr. in-4to. d. rel.   60.—
> Périodique illustré, publié pendant l'occupation allemande. On y trouve
> e.a.: Stemmen uit Göttingen. Blik in het leven onzer Vlaamsche krijgs-
> gevangenen. — Bij onze broeders in de Duitsche gevangenkampen. —
> Stemmen uit Holzminden. — etc. Contient en outre des poèmes et des
> nouvelles par Stijn Streuvels, Lode Baekelmans, A. Rodenbach, e.a.

143. — **Vries, R. W. P. de**, Oude adreskaarten. Amst. 1897. 4to. *T.
à p.*   2.—

144. — **Zur Westen, W. von**, Vom Kunstgewand der Höflichkeit. Glück-
wünsche, Besuchskarten und Familienanzeigen aus sechs Jahrhun-
derten. Berlin, 1921. Profusely illustrated. 4to. Hfvellum.   45.—
> Remarkable work for the knowledge of manners and the history of the
> art of engraving during 6 centuries, specially in Germany, Austria, Swit-
> zerland and France.

145. **Marine dictionaries. — (Aubin)**, Dictionnaire de marine conten.
les termes de la navigation et de l'architecture navale. 2e éd.
augm. Amst., J. Covens et C. Mortier, 1736. Av. front., 24 pl.
et de nombr. figg. 4to. veau.   70.—

146. — **Erdbrink, D. R.**, Engelsch vertaalboek voor zeelieden. Met
woordenboekje. Amst. 1860. 2 tom. 1 vol. cart.   2.50

147. — **Jal, A.**, Glossaire nautique. Répertoire polyglotte de termes de
marine anciens et modernes. Paris, 1848. Av. figg. 4to. peau de
truie. (1591 pp. à 2 col.).   140.—
> Ouvrage indispensable qui n'est pas encore remplacé. Très rare.

148. — **Lennep, J. van**, Zeemans-woordenboek behelz. een verklaring
der woorden bij de scheepvaart en den handel in gebruik. . . . en
der spreekwijzen daaraan ontleend. Amst. 1856. boards. *Out of
print.*   12.—

149. — **Schokker, H. W.**, Zak-woordenboek van Engelsche zeetermen,
in het Hollandsch overgebracht. 's Grav. 1841. pet. in-8vo. d.
veau.   1.—

150. **Mark Twain. — Benson, I.**, Mark Twain's Western years. Together
with hitherto unreprinted Clemens Western items. Stanford University, 1938. W. portrait and pl. cloth.   6.50

151. **Martinius, F. — Slee, J. C. van**, Franciscus Martinius, predikant
te Epe, 1638–1653. Deventer, 1904.   2.—

152. **Memorial publications. — Abhandlungen, Philosophische**, Max
Heinze zum 70. Geburtstag gewidmet von Freunden und Schülern.
Berlijn, 1906.   10.—
> Contributions de: Aall, Paul Barth, Lipps, Medicus, Raoul Richter e.a.

153. — **Album** opgedragen aan Prof. Dr. J. Vercoullie, ter gelegenheid van zijn 70en verjaardag en van zijn emeritaat. Brussel,
1927. 2 vol. Av. portrait. 4to.   7.50
> *Contient e.a.*: **C. Debaive**, Geschriften van Prof. Dr. J. Vercoullie. — **C.**

Prices are in Dutch guilders

Blancquaert, Klein-Brabantsche dialectgrenslijnen. — **P. de Reul,** L'Hellénisme de Robert Browning. — **P. De Keyser,** Wat Oud-Brussel zong in de 17e eeuw. — **A. Lodewyckx,** Duitsche nederzettingen en de Duitsche taal in Australië. — **H. Van Werveke,** Over kerkbouw in Vlaanderen vóór de 13e eeuw. — etc.

154. **Memorial publications.** — **Bioscoop-bond, Nederlandsche.** 1918
11 Febr. 1938. S. l. 1938. 4to. 1.—
Offic. orgaan, no. 61.

155. — **Célébration** du cinquantenaire du synode de l'Union des églises protestantes évangéliques de Belgique, 1889. Brux. 1890. 5.—
Les pp. 109—450 contiennent: Notices histor. sur les églises. Anvers, Gand, Maria-Hoorebeke, Courtrai-Roulers, Dour, Tournai, Verviers, Liège. Seraing, Bruxelles.

156. — **Christiansen, P.,** Hundrede aar mellem bøger. København, 1936. W. reprod. of drawings from L. Estvad, 2 of which coloured.
3.—
Charmant petit livre publ. à l'occasion du centenaire de la maison Høst & Son à Copenhague.

157. — **Feestbundel** Dr. Abraham Bredius aangeboden 18 April 1915. Amst. 1915. 2 vols. W. portr., plan and 105 pl., reprod. paintings, drawings, portrait etc. of Rembrandt, a.o. 4to. (17.50) 10.—
Contains: **F. Schmidt Degener,** Rembrandt en Homerus. — **J. W. Enschedé,** Het oude groote orgel in de Sint Bavokerk te Haarlem. — **H. E. van Gelder,** Een Haagsche fabriek van ,,Delftsch Aardewerk". — **C. J. Gonnet,** Haarlemsche glasschrijvers. — **C. Hofstede de Groot,** Rembrandts onderwijs aan zijne leerlingen. — **A. O. van Kerkwijk,** Begrafenispenningen van 17e eeuwsche Nederl. kunstschilders. — **F. Lugt,** Wandelingen met Rembrandt in Amsterdam. — etc.

158. — **Festgabe** aus Wissenschaft und Bibliothek, Otto Glauning zum 60. Geburtstag. Lpz. 1938. Vol. 2. W. portr. and ill. 4to. 5.—
Contents: **H. A. Müster,** Aufgaben der Zeitungswissenschaft und der Bibliothekswissenschaft. — **R. Oehme,** Kartograf. Bedeutung der Landkarten des Joh. Stumpf. — **W. Ronneberger,** Die Schlossbibliothek zu Jena. — **W. Schöne,** Die Inkunabeln der periodischen Presse. — etc., etc.

159. — **Festschrift** Louis Gauchat. Aarau, 1926. W. portrait, map, pl., etc. cloth. 6.50
Contains: **P. Aebischer,** La situation linguistique dans la vallée de La Roche du 13e siècle à 1500. — **A. Barth,** Beiträge zur französ. Lexikographie. — **F. Fankhauser,** Aus der Walliser Volkskunde des 18. Jahrh. — **A. Steiger,** Ursprung des spanischen Epos. — **K. Weller,** Liebe und Tod in Leopardis Gedichten. — etc.

160. — **Festschrift** zum 70. Geburtstag Otto Liebmanns. Berlin, 1910. Av. portrait. 4.75
Kant-Studien. Bd XV, H. 1.

161. — **Gedenkboek** van de Carpentier Alting-stichting, 1902—1927. Batavia, 1927. Av. pl. en couleurs et ill. 4to. 2.50

162. — **Gedenkboek** uitgeg. ter gelegenh. van het 100-jarig bestaan van het Nederlandsche Zendelinggenootschap, 1797—19 Dec. 1897. Rott. 1897. Av. plus. portr. et ill. gr. in-4to. 3.—

163. — **Gedenkboek** Billiton 1852—1927. 's-Grav. 1927. 2 vol. Av. 16 portr., cartes et pl., dont 1 en couleurs et 304 ill., dont plusieurs en grandeur de la page. gr. in-4to. toile, tête dor. (90.—) 50.—
Ce bel et intéressant ouvrage fut publié lors du 75e anniversaire de la fondation de la Société pour exploiter l'étain à Billiton.
Il est richement illustré et donne une description de l'industrie dans tout son étendu, comme elle s'est développée dans les trois quarts d'un siècle.
Pas dans le commerce.

Mart. Nijhoff, The Hague — Cat. No. 633

164. **Memorial publications.** — Geschiedenis der Nederl.-Indische Gas
Mij., hare gasfabrieken te Batavia, Soerabaja, Semarang, Buiten-
zorg en Paramaribo en hare Electrische Centrale te Batavia.
Rott. 1913. Av. portr. et pl. gr. in-4to-obl.   4.50
Pas dans le commerce.
Ajoute: **Sleyden, Ph. W. van der,** De aanslibbing van het West. en Oost.
Vaarwater van Soerabaja. 1902. *T. à p.* — **Fock, D.,** De verlegging van
den mond der Solorivier naar Sidojoe Luwas. 1902. *T. à p.*

165. — **History, Brief,** of the Netherlands Trading Society, 1824-1924.
The Hague, 1924. W. portraits, facs. and pl. roy. 4to.   10.—
The plates represent: „street in Decima", „the roadstead at Batavia
1850", „View of Paramaribo (1865)"; etc.

166. — **Jahrhunderte, Anderthalb,** Rudolf M. Rohrer, 1786-1936. Die
Geschichte einer deutschen Drucker- und Verlegersfamilie. Brünn,
1937. W. numer. portr., pl. and facs. (partly in colours). sm. fol.
cloth.   6.—

167. — **Jaren, 25,** decentralisatie in Nederl.-Indië. 1905-1930. Sa-
mensteller F. W. M. Kerchman. Batavia, 1930. W. portraits and
numer. ill. roy. 4to. hfcloth. (20.—)   10.—
H. J. Levelt, Samenstelling van de raden der autonome ressorten. — **A.
H. Stam,** De verzorging der stadshygiène door land en gemeente. — **Th.
Karsten,** Stedebouw. — De Vereeniging voor locale belangen. — Gemeente
Batavia, Semarang, Bandoeng, etc., etc. — De voormalige Gewestelijke
raden in West-Java. — etc.

168. — **Junghuhn, Franz.** Gedenkboek, 1809-1909. 's-Grav. 1910.
W. portr. and pl. (8.—)   5.—
Contains: **K. Martin,** J.'s Ansichten über versteinerungsführende Sedi-
mente von Java. — **P. van Leersum,** J.'s verdiensten voor de kina-cultuur
op Java (1856-1864). Av. 4 pl. — **S. H. Koorders** en **J. F. Niermeyer,** Over
J.'s verdiensten voor de plantengeographie van Java. — **E. C. J. Mohr,**
Over zandonderzoek. — etc., etc.

169. — **Kultur, Die,** der Abtei Reichenau. Erinnerungsschrift z. 1200en
Wiederkehr des Gründungsjahres des Inselklosters, 724-1924.
München, 1925. 2 vols. W. 2 front., 2 maps and numer. ill. cloth.
(60.—)   35.—
Contains: **K. Brandi,** Die Gründung des Klosters. — **O. Roller,** Die Mün-
zen der Reichenau. — **M. Hartig,** Die Klosterschule und ihre Männer. —
P. Lehmann, Die mittelalterl. Bibliothek. — **R. Molitor,** Die Musik der
Reichenau. — **A. Boeckler,** Die Reichenauer Buchmalerei. — etc.

170. — **Kunz, M.,** Geschichte der Blindenanstalt zu Illzach-Mülhausen
i.E. während der ersten 50 Jahre ihrer Tätigkeit, ferner deutsche,
französ. und italien. Kongreszvorträge und Abhandlungen über
das Blindenwesen. Lpz. 1907. Av. ill. gr. in-4to. toile.   5.—

171. — **Mélanges** offerts à Emile Picot. Paris, 1913. 2 forts vol. Av.
portrait, pl. et facs.   10.—
Contient un grand nombre d'articles intéressants pour la linguistique, la
littérature et les beaux-arts par Lacombe, Baudrier, Lalay, H. Cordier, De
la Roncière, Lefranc, Farinelli, etc.

172. — **Recueil** d'études, dédiées à la mémoire de N.P. Kondakov.
Archéologie. Histoire de l'art. Etudes byzantines. Prague, 1926.
Av. pl. et ill. gr. in-4to.   20.—
Contient: **V. Zlatarski,** Première campagne du Tsar Symeon de Bulgarie
contre Constantinople. (En russe). — **F. Fettich,** Die Tierkampfscene in der
Nomadenkunst. — **G. Sotériou,** Les icones sculptées en Byzance. (En grec).
— **N. Belaiew,** „Bulat" et „Karalugh". Essai sur l'histoire de l'acier en
Russie. (En russe). — **J. Cibulka,** Huminatori rukopisu Velkych Kronik

Francie. — **O. M. Dalton,** An enamelled gold reliquary. — **J. Puig i Cadafalch,** La transmission de la coupole orientale à la basilique romane. — **L. Niederle,** Keramika západnich skythskych mohyl. — etc.

173. **Memorial publications.** — **Ruprecht, W.,** Väter und Söhne. Zwei Jahrhunderte Buchhändler in einer deutschen Universitätstadt. Göttingen, Vandenhoeck & Ruprecht, 1935. W. pl., contain. 24 ill. cloth. 3.—

174. — **Schmitz, H.,** Berliner Eisenkunstgusz. München, 1917. W. front., 44 pl., represent. statuettes, medals, objects of art, etc. roy. 4to. boards. (15.—) 6.—
   „Festschrift z. 50 j. Bestehen des Kgl. Kunstgewerbemuseums, 1867-1917."

175. — **Sluyterman, K.,** Gedenkboek van de Vereeniging „Arti et industriae" tot bevordering der kunstnijverheid. Uitgeg. n. aanleiding van het 25-jarig bestaan der Vereeniging, 1884-1909. 's-Grav. 1909. W. 131 ill. 4to. cloth. (10.—) 3.—

176. — **Staat und Persönlichkeit.** Erich Brandenburg zum 60. Geburtstag dargebracht von A. Doren, P. Kirn, J. Kühn, P. Ostwald, W. Stach, H. Wendorf, u.A. Lpz. 1928. W. portr. cloth. (9.—) 5.—
   Contents: **W. Stach,** Zu Cäsars Nachrichten über den Ackerbau bei Sueben u. Germanen. — **G. Vrind,** Cäsar und seine Kommentare. — **K. Weimann,** Landesherr und Allmende. **H. Schulz,** Das Recht zur Revolution. — etc., etc.

177. **Monroe.** — **Madisson, J.,** Manifeste du gouvernement américain, (10 févr. 1815), ou causes et caractère de la dernière guerre d'Amérique avec l'Angleterre. Trad. sur la 11e éd. angl. p. (Ch. Malo). Paris, 1816. cart. 12.50
   Cet opuscule est aussi attribué à Monroe. Il est de premier intérêt pour l'histoire politique des Etats-Unis et la doctrine de Monroe.

178. **Ontcommer.** — **Gessler, J.,** De Vlaamsche baardheilige Wilgefortis of Ontcommer. Antw. 1937. W. 52 ill. 2.80
   This monography is based upon many unknown documents, especially relating to Northern France, which throw a new light upon the origin of the legend. The illustrations were for the greater part unpublished.

179. **Ovidius,** Metamorphoseon ll. XV. C. annot. J. Min-ellii, suppl. et emend. P. Rabus. Amst., C. Fritsch et M. Böhm, 1710. pet. in-8vo. vél. 2.—

180. — Der Griecxser princerssen, ende jonckvrouwen clachtige sendt brieven, heroidum epistolae ghenaempt. Nu eerst in Duytsche duer Cornelis van Ghistele rhetorijckelijck overgheset. Thantwerpen, Hans de Laet, 1559. sm. 8vo. gilt vellum. 90.—
   First edition of this translation.
   Fine copy with the initials of Moeritgen Pietersson, whose signature is on the titlepage, on the binding, and dated: 1617.

181. **Palaeography.** — **Astle, Th.,** The origin and progress of writing, as well hieroglyphic as elementary, ill. by engravings taken from marbles, manuscripts and charters, ancient and modern; also some account of the origin and progress of printing. 2d ed. W. add. London, 1803. Av. portr. et pl., dont qq.-unes color. fol. cart., n. r. 25.—

182. **Palaeography. — Bastard, A. de,** Une collection de 55 planches tirées de la première partie (*Paléographie*) de l'ouvrage somptueux sur les peintures et ornements des manuscrits. Paris, 1832–69.

                                                                    300.—
Ces planches reproduisent des Ecritures mérovingiennes, 1e à 3e époque (6e à 8e sc.). — Écritures carlovingiennes, 1e à 2e époque, visigothiques, lombardiques, franco-saxonnes (8e à 9e sc.).

Presque toutes ces planches ont été exécutées en couleurs et en or, et en pourpre et en or et elles représentent de superbes lettres remarquables d'un gout le plus pur, comme: écriture dite chrysographie, lettres et initiales dites zoomorphes, ornithomorphes, initiales en broderie dites ailées et fleuronnées, initiales dites serpentines, initiales à figures d'hommes, alphabét. ichthyomorphe, initiales phyllomorphes, initiales dites perlées, entrelacées, brodées, écriture capitale dite en marqueterie.

Comme un ex. complet (140 pl.) est des dernières raretés et se vend à 2000 florins environ, une collection comme la nôtre doit être très appréciable pour toute institution qui s'intéresse à l'ancienne paléographie.

183. **— Brugmans, H.,** en **O. Oppermann,** Atlas der Nederlandsche palaeographie. 's-Grav. 1910. Av. 28 pl., conten. 56 reprod. de facs. et texte explic. sur papier de Hollande. fol. En portef. (30.—)
                                                                    18.—
Les planches représentent des facs. de lettres, chartes, contrats, e.a. documents officiels dès l'année 1100 jusqu'à 1692.

184. **— Chassant, A.,** Paléographie des chartes et des manuscrits du XIe—XVIIe siècle. Evreux, 1839. Av. 8 pl.   5.—

185. **— Chatelain, E.,** Paléographie des classiques latins. Paris, 1884, 1900. 2 vol. Av. 210 pl. de facs. gr. in-fol. En livr.   80.—
Importante publication, contenant des fascimiles de fragments des plus anciens mss. connus de Plaute, Térence, Varron, Catulle, Cicéron, Virgile, Ovide, Juvénal, Pétrone, Lucain, Sénèque, Suétone, e.a.
Epuisé.

186. **— Holle, K. F.,** Tabel van Oud- en Nieuw-Indische alphabetten. Bijdr. tot de palaeographie van Nederl.-Indië. Uitgeg. d. h. Bat. Genootsch. v. K. en W. Batavia, 1882.   5.—
Très estimé.

187. **— Koning, J.,** Verklaring van het oud letterschrift. Leyden, 1818. pet. in-8vo. Av. atlas de 17 pl. 4to.   10.—
Epuisé et encore recherché.

188. **— Paléographie musicale.** Fascimilés phototypiques des principaux manuscrits de chant grégorien, ambrosien, mozarabe, gallican, publ. par les Bénédictins de l'abbaye de Solesmes. Tournay, 1889–1934. T. I–XIII, XIV, nos. 1–16 (= nos. 1–155). 14 vol. — Idem. 2e Série (monumentale). Solesmes, 1900–24. T. I, II. 2 vol. — Ens. 16 vol. 4to. dont la 1re série en 14 vol. d. mar. rouge du Levant, non rogné, tête dor., les deux autres vol. d. veau et br.   1550.—
Superbe exemplaire complet, excessivement rare.

189. **— Paoli, C.,** Grundriss zu Vorlesungen üb. Latein. Palaeographie u. Urkundenlehre. A. d. Ital. von K. Lomeier. 2e Aufl. Innsbruck, 1889–1900. 3 vol. toile et br.   4.—

190. **— Silvestre, J. B.,** Paléographie universelle. Collection de fascimile d'écritures de tous les peuples et de tous les temps. Accomp. d'explications histor. et descript. p. Champollion-Figeac et A. Champollion fils. Paris, 1841. 4 vol. Av. de nombr. superbes pl. gr. in-fol.

Plein mar. rouge, richement doré, dor. s. tr. (*Lewis*). *Très bel ex.*
300.—

L'édition originale française de l'ouvrage célèbre de Silvestre, de beaucoup supérieure à l'édition anglaise, quant à l'exécution des planches, qui ont été *coloriées* avec le plus grand soin et rehaussées d'or.

191. **Palaeography.** — **Wattenbach, W.,** Anleitung z. griechischen Palaeographie. 2e Aufl. Lpz. 1877. 4to. Av. 12 tables d'écritures. pet. in-fol. 2.50

192. **Peutinger, C.** — **Herberger, Th.,** Conrad Peutinger in seinem Verhältnisse zu Maximilian I. Augsburg, 1851. Av. portr. 4to. 1.50

193. **Philosophy.** — **L'Année philosophique.** 1867, 1868. Etudes critiques sur le mouvement des idées générales p. F. Pillon. Paris, 1868, 69. Année 1, 2. — (*Continuée par:*) **La Critique philosophique, politique, littéraire.** Publ. p. Ch. Renouvier. 1872–89. Année 1—13; N. Série, année 1—5. 18 tom. 35 (sur 36) vol. Av. index de la 1re série. — (*Continuée par:*) **L'Année philosophique,** 1890–1913. Publ. p. F. Pillon. 1891–1914. Année 1—24. 24 tom. 12 vol. — Ens. 63 tom. 51 vol., dont 14 d. rel. 250.—

Collection complète, comme on ne rencontre presque jamais. Les premières années de l'Année philosophique et la Critique philosophique surtout sont d'une grande rareté.

194. — **Aristoteles,** Opera. Ed. Academia Regia Borussica. Ex recogn. I. Bekkeri e. a. Berol. 1831–70. 5 vols. 4to. vellum. *Fine copy.*
175.—

The original impression.

195. — **Bauer, E.,** Das Urchristentum, das ist Christi Lehre in ihrer ursprüngl. Reinheit. S. l. 1845. cart. 7.50

196. — **Buhle, J. G.,** Geschichte der neuern Philosophie. Göttingen, 1800–05. 6 tom. 9 vol. cart. 18.—

197. — **Clenardus, Nic.,** Epistolarum ll. II. Antv., Chr. Plantin, 1566. Full red mor., gilt edges, fly-leaves of blue silk, dent. and fil. on covers, dent. intér. *Fine copy.* 75.—

The Fleming Clenardus or Kleinarts was professor in Spain. He devoted himself particularly to the study of the Arabian language.
The most complete and very scarce edition of his „Epistolae". It was publ. by Clusius, the celebrated botanist. The „liber I" contains letters already publ. in 1550, 1551 and 1561, the letters of „liber II" were brought from Spain by Clusius and were publ. here for the first time. See Hunger, Charles de l'Écluse. La Haye, 1927.
Bound up with:
De optimo genere disputandi colloquendique ad Ianum Gontaldum Bironem. Paris., G. Morelius, 1551 (at the end 1552).

198. — **Défenseur, Le,** des droits de l'homme. Journal de propagande politique. Prospectus-specimen (signé A. C.(omte?). Paris, 1833. fol. 2.50

Hatin, p. 388. Seul no. paru.

199. — **Descartes,** Œuvres. Publ. p. Ch. Adam et P. Tannery. Paris, 1897–1913. 12 vol. Av. portr., pl. et figg. et index. Ens. 13 tom. 12 vol. 4to. d. rel. 225.—

Importante publication, éditée d'une façon magistrale avec des caractères du XVIIe siècle.

200. — **Erdmann, J. E.,** Grundriss der Geschichte der Philosophie. 4e (letzte) Aufl. bearb. von B. Erdmann. Berlin, 1896. 2 vol. d. veau. 24.—

20 PHILOSOPHY

201. **Philosophy.** — **Fichte, J. G.**, Sämtliche Werke. Berlin, 1845–46. 8 vols. Hfcalf. (Bindings somewhat worn). 60.—
Original imprint.

202. — **Fries, J. F.**, Neue Kritik der Vernunft. Heidelberg, 1807. 3 vol. d. veau. 40.—
Edition originale, très rare.

203. — — System der Metaphysik. Heidelberg, 1824. cart. 15.—

204. — **Gauchat,** (L'abbé), Lettres critiques ou analyse et réfutations de divers écrits modernes contre la religion. Paris, 1758–63. 19 vol. pet. in-8vo. veau. 75.—
Périodique, consacré à la réfutation des doctrines des encyclopédistes. Bayle, Pope, Diderot, Montesquieu, Voltaire, Rousseau, Holbach, Helvetius, e.a. y sont violemment attaqués. Le t. XVI est entièrement consacré à la réfutation du ,,monstrueux Code de la Nature de Morelly.'' Tout ce qui a paru.

205. — **Goldscheid, R.**, Höherentwicklung und Menschenökonomie. Grundlegung der Sozialbiologie. Lpz. 1911. T. I (seul paru). toile. *Epuisé*. 5.—
Philosoph.-soziolog. Bücherei. VIII.

206. — **Goslinga, W. J.**, De rechten van den mensch en burger. Overzicht der Nederlandsche geschriften en verklaringen. 's-Grav. 1936. 2.—

207. — **Johannes von Lykopolos**, Ein Dialog über die Seele und die Affekte des Menschen. Hrsg. von S. Dedering. Uppsala, 1936. 5.—
Contains the Syrian text with introd., variants, résumé of the contents, etc.
Arbeten utg. av V. Ekmans universitetsfond, Uppsala. Nr. 43.

208. — **Klaiber, J.**, Hölderlin, Hegel und Schelling in ihren schwäbischen Jugendjahren. Stuttgart, 1877. rel. 5.—

209. — **Lehodey, V.**, Le saint abandon. 3e éd. Paris, 1921. 2.—
Nature, fondement, objet et excellence du saint abandon.

210. — **Lichtenberger, F.**, Histoire des idées religieuses en Allemagne depuis le XVIIIe siècle jusqu'à nos jours. 2e éd. Paris, 1888. 3 vol. 5.—

211. — **Masaryk, Th.**, Zur russischen Geschichts- und Religionsphilosophie. Soziologische Skizzen. Jena, 1913. 2 vols. cloth. 30.—

212. — **Mather, Jr., F. J.**, Concerning beauty. Princeton, 1935. W. front. and 10 reprod. of paintings. cloth. 6.—

213. — **Mendelssohn, M.**, Gesammelte Schriften. Nach den Originalen und Hss. hrsg. von G. B. Mendelssohn. Lpz. 1843–45. 7 vols. W. portrait. Hfmor. 30.—
Best edition.

214. — **Novissimum Organon, Le.** Organe instructeur de l'enseignement mutuel social populaire. Revue trimestrielle, réd. par l'Ecole du ,,Hieron''. Paray, 1895–1900. 6 vol. Av. quelques figg. — (*Continué par:*) **Le Politicon** pour l'instruction supérieure diplomatique. Suivant les règles et disciplines du Sacré-Coeur. Paray, 1901–06. 6 vol. Av. cartes en couleurs. — (*Continué par:*) **Le Pam-Epopéion.** Annales de l'école bardique et de l'école diplomatique internationales. Publ. sous les auspices de l'Aréopagie Auréolée Mariale. Réd. A. de Sarachaga. Paray, 1907–11. 4 vol. — Ens. 16 vol. gr. in-4to. 75.—

Prices are in Dutch guilders

215. **Philosophy.** — **Origenes,** Opera omnia, ed. C. et C. V. Delarue, denuo rec. C. H. E. Lommatzsch. Berlin, 1831–48. 25 vols. bound. 70.—
     The best edition.

216. — **Perry, R. B., A. Ch. Krey, E. Panofsky a.o.,** The meaning of the humanities. Five essays. Princeton, 1938. cloth. 5.—
     History of art as a humanistic discipline. — Theology and the humanities. — Literature and the humanities. — etc.

217. — **Pfleiderer, O.,** Geschichte der Religionsphilosophie von Spinoza bis auf die Gegenwart. 3e (letzte) Aufl. Berlin, 1893. d. rel. et br. *Très rare.* 25.—

218. — **Prihonsky, F.,** Neuer Anti-Kant oder Prüfung der Kritik der reinen Vernunft nach den in Bolzano's Wissenschaftslehre niedergelegten Begriffen. Bautzen, 1850. sm. 8vo. hfcloth. 40.—
     Excessively rare work against Kant, of which no other edition has appeared.

219. — **Schlick, M.,** Gesammelte Aufsätze. Wien, 1938. bound. 12.50
     Moritz Schlick, being one of the leading men of the „Wiener Kreis" has been of great influence on the development of philosophical thought. This volume contains his production of the last 10 years (1926–1936).

220. — **Schotel, G. D. J.,** A. M. van Schurman. M. aanteek. en bijl. 's-Hert. 1853. Av. portr. et facs. cart. 3.50

221. — **Spinoza,** Tractatus theologico-politicus. Hamburg, (Amst., C. Conradus), 1670. 4to. vellum. 250.—
     The real „Editio princeps". In the year 1670 four editions have been publ., but only one of them was supervised by Spinoza himself. In the editions B, C and D there are numerous mistakes; it is the edition A of which Land and van Vloten have made use, when they published the „Opera" in 1882. The edition A is of the utmost rarity.

222. — — De nagelate schriften, als zedekunst, staatkunde, verbetering van 't verstant, brieven en antwoorden. Vert. (d. Iarig Iellis). (Amst.) 1677. W. fine portrait. 4to. orig. vellum. (Back slightly damaged). 200.—
     Van der Linde, nr. 23. T h e e x t r e m e l y r a r e d u t c h e d i t i o n. The translation is by J. Hz. Glazemaker, and is publ. by J. Rieuwertsz at Amsterdam, at the same time as the latin edition. The dutch edition is important, because the text has been translated from the ms. itself and therefore enables the correction of doubtful or wrong passages of the latin edition.
     Copies with the portrait are excessively rare.

223. — **Stahl, F. J.,** Die Philosophie des Rechts. 5e (letzte) Aufl. Tübingen, 1878. 3 vol. 30.—
     I. Geschichte der Rechtsphilosophie. — II—III. Rechts- und Staatslehre auf der Grundlage christlicher Weltanschauung.
     Impression originale de la dernière édition. Fort rare.

224. — **Steffens, H.,** Caricaturen des Heiligsten. Lpz. 1819. 2 vol. d. veau. 12.—
     Steffens, né à Stavanger en 1773, partisan de la doctrine de Spinoza et ami de Schelling, Schlegel, Tieck e.a., se fit un nom tant au sujet de l'histoire naturelle (minéralogie) et de la philosophie naturelle, que par ses poèmes et ses romans. Il s'occupa aussi des questions politiques, surtout au temps de la chute de Napoléon, du Congrès de Vienne, etc., quand il fit paraître e.a. son „Caricaturen des Heiligsten". Voir Allgem. Deutsche Biographie, t. XXXV, pp. 555—558.

225. — **Tennemann, W. G.,** Geschichte der Philosophie. Lpz. 1798–1819. 11 tom. 12 vol. d. veau. 30.

226. **Philosophy.** — (**Vollgraff, C. F.**), Polignosie und Polilogie oder: genetische und comparative Staats- und Rechts- Philosophie auf anthropognost., ethnolog. und histor. Grundlage. Marburg, 1855. cart. (1013 pp.).   18.—
     Versuch einer wissenschaftl. Begründung der allgem. Ethnologie etc. Bd III.

227. — **Wahl, J.**, Etudes Kierkegaardiennes. Paris, 1938. toile. (747 pp.).
                                                                          5.—
     Collection „Philosophie de l'esprit."

228. — **Weiss, P.**, Reality. Princeton, 1938. cloth.   7.—
     „This book is an important original contribution to systematic epistolog-ical and metaphysical theory."

229. — **Willm, J.**, Histoire de la philosophie allemande depuis Kant jusqu'à Hegel. Paris, 1846–49. 4 vol. d. veau.   22.50

230. **Polyglot bible.** — **Biblia Regia**, hebraice, chaldaice, graece et latine; cura et studio B. Ariae Montani. Antwerpiae, Plantin, 1569–73. 8 vols. W. maps and pl. by P. van der Heyden, J. Wiericx, Ph. Galle, a.o. fol. old calf, with coats of arms.   600.—
     The celebrated Polyglot Bible of Plantin is one of the most famous typographical monuments. The 8th volume is of special interest for the highly remarkable world map by Montanus. From this mappemonde the „Terra Australis Incognita" has disappeared but a small part of Northern Australia has been indicated. The cartography of Australia originated with this map.
     The bindings are faded and some backs are repaired, the lower margins of one vol. somewhat waterstained, but otherwise a very fine, clean and very tall copy, meas. 42 × 28 c.M.
     See Ruelens en de Backer, Annales Plantiniennes and Rooses, Le Musée Plantin-Moretus, 1914, pp. 71–97.

231. **Portugal: History, Literature, Arts.** — **Alma nacional.** Revista republicana. Dir. A. J. d'Almeida. Lisboa, 10 de fevr. — 29 de set. 1910. 34 nos. W. some caricat. pl. Bound in 1 vol. 4to. hfcalf.
                                                                         12.—
     Collaborators: J. de Freitas, Th. da Fonseca, A. Vaz, G. Lima, a.o. All published.

232. — **Annaes das sciencias, das artes, e das letras**; por huma sociedade de Portuguezes residentes em Paris. Paris, 1818–22. 16 vol. d. veau.   48.—
     Qq. trous de vers dans les reliures.

232a.— **Archeologo português, O.** Collecção illustr. de materiaes e noti-cias, publ. p. Museu ethnographico português. Lisboa, 1895–1933. T. 1–29. 29 tom. en 15 vol. Av. cartes, pl. et ill. d. veau unif.   175.—

233. — **Arquivo literario.** Dir. Delfim Guimarães. Lisboa, 1922–29. 4 vol. d. veau.   30.—
     Journal littéraire très-important, contenant des articles, poèmes, etc. par D. Bernardez, L. de Mendonça, A Pimentel, C. de Passos, J. Machado, e.a.
     Tout ce qui a paru.

234. — **Artes e Letras.** Revista de Portugal e Brazil. Lisboa, 1872–75. T. 1–3, 4, nos. 1–5. 4 vol. Av. 82 pl. et de nombr. ill. gr. in-4to.
                                                                         30.—
     Tout ce qui a paru. Les tom. I et III sans le titre et la table.

235. — **Atlantida.** Mensario artistico, literario e social para Portugal e Brazil, (later:) **Atlantida.** Orgão do pensamento latino no Brazil e em Portugal. Lisboa, 1915–19. 4 years (= 12 vols. = 48 nos.).

           Prices are in Dutch guilders

W. pl. and ill. by R. Bernardeli, R. Lino, Santos e Silva, a.o.
Bound in 7 vols. hfcalf, covers preserved. 60.—
Collaborators: M. Sousa Pinto, T. de Queiroz, A. Gil, J. Figueiredo,
J. Cortesão, a.o.
All published.

236. **Portugal: History, Literature, Arts.** — **Cosmorama litterario, 0.**
Jornal da Sociedade escholastico-philomatica. Lisboa, 4 de jan.–
agosto 1840. 33 nos. *Tout ce qui a paru.* — **0 Corsario.** Jornal
de litteratura e recreio. Lisboa, 2 de abril–7 de maio 1838. 5 nos.
Av. 5 pl. *Tout ce qui a paru.* — **0 Viajante.** Jornal recreativo
semanal. Lisboa, 13 de out. 1838–12 de jan. 1839. 14 nos. *Tout
ce qui a paru.* — Rel. en 1 vol. 4to. (Rel. endomm.). 12.50

237. — **Epoca, A.** Jornal de industria, sciencias, litteratura e bellas
artes. Lisboa, 1848, 49. 2 vol. (= 52 nos.). Av. ill. Rel. en 1 vol.
4to. d. veau. 18.—
Journal très important conten. des articles, des romans, etc. par Rebelo
da Silva, Lopes de Mendonça, Andrade Corvo, e.a.
Tout ce qui a paru.

238. — **Espectro de Juvenal, 0.** Red. G. Leal, G. d'Azevedo, L. Cor-
deiro, M. Lima, S. Pinto. Lisboa, 1872–73. 3 nos. 7.50
Tout ce qui a paru.

239. — **Instructor Portuense, 0.** Periodico mensal, contendo artigos
de educaçao, litteratura, moral, historia, sciencias e artes. Trad.
de varias linguas. Porto, 1844. 1 vol. Av. 12 pl. de J. F. Ribeiro.
d. veau. 10.—
Contient: Curiosidades naturaes do Mexico. — Naçoes indigenas de
Sancta-Catharina, no Brasil. — Primeiros estabelecimentos na Pennsyl-
vania. — Convenção de Guilherme Penn com os Indios delawares. —
Fundaçao de Philadelphia. — Chili (Zoologia). — etc., etc.
Tout ce qui a paru.

240. — **Jornal de bellas-artes.** Red. F. de Sequeira Barreto; R. Paga-
nino. Lisboa, 1857. 8 nos. W. numer pl., original etchings, engravings and woodcuts. Bound in 1 vol., covers preserved. (Back
damaged). 20.—
All published.

241. — **Lanterna, A.** Folha politica. Lisboa, 1869–73. 10 vols. hfcloth.
60.—
Feuille hebdomadaire, bi-hebdom. et mensuelle, contenant de véhé-
ments articles polémiques, politiques et révolutionnaires. Elle a paru
sous des titres très variés, e.a.: A Lanterna. — A luz de lanterna. —
A lanterna e.a. luz. — O precursor. — O clarim. — O clarao. — O clamor.
— O facho. — O pharol. — A aurora. — A estrella. — A fama. — ....A
camarilha. — A bancarota. — .... O anno bon. — .... A falla do throno.
— O imposto. — Os Bonapartes. — Napoleao III. — .... A propaganda.
— O raio.
Complete collection. See da Silva, Diccionario, t. XIII, p. 280.

242. — **Memorias de litteratura Portugueza,** publ. p. Academia Real
das Sciencias de Lisboa. Lisboa, 1792–1812. 8 vols. hfcalf. 60.—
Complete collection. All vols. of the original imprint, except vol. 8,
which is of the reprint of 1856.

243. — **Museu Portuense.** Jornal de historia, artes, sciencias, industriaes
e bellas letras. Publ. debaixo dos auspicios da sociedade da typo-
graphia commercial portuense. Porto, 1839. 12 nos. En 1 vol.
Av. grav. s. bois. pet. in-fol. d. veau. 12.—
Tout ce qui a paru.

24 PORTUGAL

244. **Portugal: History, Literature, Arts.** — **Panorama, O.** Jornal litterario e instructivo da Sociedade propagadora dos conhecimentos uteis. Lisboa, 1837–68. 18 vol. Av. ill. Rel. en 9 vol. d. veau. **85.—**
Périodique très estimé auquel ont collaboré les meilleurs auteurs du temps comme A. Herculano, Garrett, Rebelo de Silva,. C. Castelo Branco e.a.
Tout ce qui a paru.

245. — **Semana litteraria, A.** Lisboa, 14 de abril–5 de maio 1889. 4 nos. en 1 vol. d. veau. **7.50**
Collaborateurs: F. d'Almeida, S. Pinto, L. Osorio, etc.
Tout ce qui a paru.
Dans la même reliure: 1. **Album comico.** Collecçao de anecdotas, epigrammas, satyras. Lisboa, 1877. Av. ill. humor. — 2. **J. Palhais,** Os dois avarentos. Lisboa, 1887.

246. — **Amorim, Fr. Gomes de,** Garrett. Memorias biograph. Lisboa, 1881–84. 3 vol. Av. portrait et facs. veau. *Bel ex.* **12.50**

247. — **Arthur, R.,** Arte e artistas contemporaneos. Lisboa, 1896, 98. Series, I, II. 2 vols. W. portr. and ill. hfcalf. **6.50**

248. — **Brito, B. de,** Monarchia Lusytana que contem as historias de Portugal (—1423). Lisboa, 1683–1752. 8 vols. fol. calf. **250.—**
Partes I et II. Lisboa, 1690. Partes III e IV por Ant. Brandao. Lisboa, 1690–1725. — Partes V et VI por Ant. Brandao. Ib. 1751–1752. — Pars VII por Rafael de Jesus. Ib. 1683. — Pars VIII por Manoel dos Santos. Ib. 1752.

249. — **Cacegas, L.,** Historia de S. Domingos, partic. do reyno e conquistas de Portugal. Reform. em estilo e ordem ampl. e particul. p. L. de Sousa. (2a ed.). Lisboa, 1767. 4 vols. W. 4 engraved titlepages. fol. calf. **150.—**

250. — **Camões, L. de,** Os Lusiadas. Poema epico. C. vida d'este poeta.... muitas notas p. J. da Fonseca. Paris, 1846. Av. portr. **7.50**
,,E' ediçao estimada e os exemplares não vulgares." (Pinto de Mattos).

251. — — Même ouvrage. Ed. critica commemor. do terceiro centenario da morte do poeta. Pupl. p. E. Biel. Lpz. 1880. Av. 2 portr., des sous-titres en couleurs et de nombr. pl., grav. s. acier d'après Begas, Kostka, e.a. gr. in-4to. mar. vert orné. **45.—**
Edition de luxe.

252. — **Cancioneiro** d'elrei D. Diniz. Pela 1ra vez impresso s. o manuscripto da Vaticana, c. alg. notas ill. p. C. Lopes de Moura. Pariz, 1847. W. 1 facs. hfmor. *Very scarce.* **28.—**

253. — **Cancioneiro** d'Evora. Publ. av. notice littéraire-histor. p. V. E. Hardung. Lisboa, 1875. d. veau. **6.—**
,,Le Cancioneiro d'Evora appartient à la fin du 16e siècle et fut probablement composé entre 1590 et 1600."

254. — **Cancioneiro Portuguez** da Vaticana. Ed. critica.... acomp de um glossario e de uma introd. s. os trovadores e cancioneiros Portuguezes por Th. Braga. Lisboa, 1878. 4to. Hfcalf. **40.—**

255. — **Chaves, L.,** Portugal àlém. Notas etnográficas. Gaia, 1932. T. I. Av. pl. **2.50**

256. — **Duijl, A. G. C. van,** Tien dagen in Portugal. 's-Grav. 1884. pet. in-8vo. **1.—**

257. — **Figueiredo, A. C. Borges de,** Coimbra antiga e moderna. Lisboa, 1886. Av. plan et 2 pl., dont 1 en couleurs. d. veau. **6.—**

Prices are in Dutch guilders

258. **Portugal: History, Literature, Arts. — Macedo, D. R.**, Obras. Lisboa, 1743. 2 tom. 1 vol. 4to. veau. 10.—
Contient e.a.: Juizo histor., jurid., e polit. s. a paz, celebr. entre as coroas de França, e Castella, 1660. — Panegyrico histor. genealog. da caza de Nemurs. — Nascimento, e genealogia do Conde D. Henrique de Portugal. — Discursos politicos, e obras metricas. — **de Balsac**, Aristippo, ou homem de Corte.
Une piqûre.

259. — **Oliveira, A. de**, Descripçam corografica do reyno de Portugal. Lisboa, 1755. Av. 7 cartes. 4to. veau. 12.—
Da Silva ne mentionne pas de cartes.

260. — **Osorio, A. de Castro**, As mulheres Portuguêsas. Lisboa, 1905. 1.50

261. — **Osorius, H.**, De rebus Emmanuelis regis Lusitaniae.... gestis ll. XII. Olysippone, Ant. Gondisalvus, 1571. W. engraved titlepage. fol. hfcalf. 50.—
Original edition.
Titlepage damaged and mounted; the 2 last leaves damaged.

262. — **Ramos-Coelho, J.**, Historia do Infante D. Duarte (1605–1649), irmão de el-rei D. João IV. Lisboa, 1889, 90. 2 stout vols. W. portr., facs. and pl. 12.—
D. Duarte visited as ambassador Spain, Savoia, Toscane, Mantua, Modena, Parma, Milan, Vienna, Stuttgart, Holland, England, Sweden, etc.

263. — **Resende, G. de**, Chronica dos valerosos, e insignes feytos del rey Dom Joam II. Lisboa, 1752. fol. calf. 18.—
Titlepage repaired and stained, otherwise in good condition.

264. — — Cancioneiro geral. Nova ed. p. A. J. Gonçalvez Guimarãis. Coimbra, 1910–17. 5 vol. d. veau. 45.—
Ce „Cancioneiro" contient les poésies d'un grand nombre d'écrivains portugais de la fin du 15e siècle. Parmi ces pièces il s'en trouve en langue espagnole, mais composées par des Portugais. De l'édition originale, parue en 1516, on ne connait que 2 exx. complets.
Joias literarias. II.

265. — **Rezende, de**, Elogio histor. de José de Seabra da Silva. Lisboa, 1861. Av. portr. 4to. 2.—
Etude très documentée sur cet homme d'état portugais du 18e siècle.

266. — **Romanceiro** e cancioneiro do Algarve. Accomp. de importantes notes por F. X. d'Athaide Oliveira. Porto, 1905. orig. boards. *Scarce.* 15.—

267. — **Santos, Cl. J. dos**, e **B. de S. Clemente**, Estatisticas e biographias parlamentares portuguezas. 1821–1892. Publ. no Jornal „O commercio do Porto". Porto, 1887–92. 3 tom. 6 vol. Av. portr. et facs. d. veau. 20.—

268. — **Satyricos Portuguezes.** Collecção de poemas heroi-comico-satyricos. Nova ed. c. introd. e anot. de J. Ribeiro. Rio de Janeiro, 1910. d. veau. 7.50

269. — **Soares da Silva, J.**, Gazeta em forma de carta, 1701–1716. Lisboa, 1933. Vol. I (1701–1703). W. 3 pl. of facs. roy.8vo. 3.—
„O texto.... é sempre curiosíssimo pela luz que jorra sôbre a vida social da Lisboa setecentista e, não menos, pelas informacões minuciosas que da para a história.... da guerra da sucessão."

270. — **Soriano S. J. da Luz**, Historia do cerco do Porto. Preced. de uma noticia s. as differ. phazes politicas da monarchia desde os

mais antigos tempos. Lisboa, 1846, 49. 2 vols. W. map. calf.
12.50
Original edition.
271. **Portugal: History, Literature, Arts. — Sousa, A. C. de,** Historia
genealogica de casa real portugueza desde a sua origem até o
presente, com as familias ilustres. Lisboa, 1735–49. 12 vols. in
14. — Provas de Historia geneal. da casa real portugueza, ti-
rados dos instrumentos dos archivos de Torré do Tombo. Lisboa,
1739–48. 6 vols. Tog. 20 vols. 4to. calf, gilt back, gilt edges. 600.—
Fine complete set of this valuable work of the greatest scarcity. Vol. 7
slightly waterstained and name on titlepages.
272. **— Teixeira, M.,** Poesias e poemas. Penumbras. — Idyllio de
Theocrito. — Occantico dos canticos. 2a ed. Rio de Janeiro,
1888. Av. portrait.   3.—
273. **Raskol, Le.** Essai historique et critique sur les sectes religieuses
en Russie. Paris, 1859. d. rel.   25.—
De grande rareté.
274. **Reden, von,** Vergleichende Kultur-Statistik der Gebiets- und Be-
völkerungsverhältnisse der Gross-Staaten Europa's. Berlin, 1848.
cart.   7.50
275. **Refugees. — Verdeyen, R. W. R.,** Belgie in Nederland, 1914–1919.
De vluchtoorden Hontenisse en Uden. 's-Grav. 1920. Av. pl.
toile. (7.50)   3.—
Description économique de la vie quotidienne des réfugiés belges dans
les refuges de Hontenisse et Uden, Pays-Bas, pendant la guerre européenne.
276. **Reize, De avantuurlyke,** of de post-koets in den modder. U. h.
Fr. Leiden, A. Ambrullaart, 1730. Av. front. et 4 pl. pet. in-8vo.
cart.   10.—
277. **Réponse** au Discours d'un soi-disant bon Hollandais, sur la liberté
de porter des munitions navales en France, etc. par un négociant
d'Amsterdam. Leide, C. de Pecker, 1779.   1.50
278. **Ridgway, J. L.,** Scientific illustration. Stanford University, 1938.
W. coloured front., pl. and figg. cloth.   8. —
,,This is a practical working manual of the principles and best practices
in the selection of appropriate scientific illustrative material methods ot
preparation, proper fitting.... and reproduction for illustrating scientific
publications.''
279. **Roman-Catholicism. — Carové, F. W.,** Ueber alleinseligmachende
Kirche. Frankfurt a.M. 1826. cart.   5.—
280. **— —** Die letzten Dinge des römischen Katholicismus in Deutschland. Lpz. 1832. d. veau.   5.—
281. **— —** Römischer Katholizismus in der Pabststadt u. Metropolen
Italiens. Lpz. 1851.   3.—
282. **— Habets, J.,** Geschiedenis van het tegenwoordig bisdom Roermond en van de bisdommen ,die het in deze gewesten zijn vooraf-
gegaan. Roermond, 1875–92. 3 vol. d. rel.   25.—
Epuisé. D'un 4e tome seulement les pp. 1 à 208 ont été publiées. Ce tome
est de toute rareté.
283. **— Swaving, J. G.,** Galerij van Roomsche beelden, of beeldendienst
der XIX eeuw. Dordrecht, 1824. — **Id.,** Roomsche feest- en heilige
dagen, of verbijstering van het menschelijke verstand. Dordrecht,

1825. — Ens. 2 vol. Av. 3 pl., dont une (en gravure) représente une procession à Montaigu (Scherpenheuvel). d. veau. 3.—

284. **Roman-Catholicism. — Swaving, J. G.**, Mêmes ouvrages, le premier en 2e éd. Dordrecht, 1841. — Ens. 2 vol. Av. 3 pl. d. rel. 3.—
La pl. de Scherpenheuvel lith.
— See also: **Ontcommer.**

285. **Rosicrucians. — Wittemans, Fr.**, Geschiedenis van de Orde der Rozekruisers. 's-Grav. 1921. 1.—

286. **(Rothe, J.)**, Een nieuwe hemel en aerde. Het nieuwe Jerusalem. De wederoprechtinge aller dingen, volgens (Act. 3. 21.), etc. 2e dr. verm. Amst., gedruckt voor den autheur, 1673. Av. 3 grav. 4to. 10.—
L'auteur était un fanatique religieux, qui se croyait prédestiné à assister le Christ à la fondation du règne millénaire.

287. **Rourke, C.**, American humor. Study of the national character. N. York, 1931. cloth. (8.75) 2.50

287a. **Rijnsburg. — Hüffer, Maria**, Die Reformen in der Abtei Rijnsburg im 15. Jahrhundert. Münster i. W. 1937. 5.- -
Vorreformationsgeschichtl. Forschungen. Bd. XIII.

288. **— (Oudaan, J.)**, Aanmerkingen over het verhaal van het eerste begin en opkomen der Rijnsburgers. 2e dr. verm. met een bij-voegsel. Rott., I. Naeranus, 1672. 4to. br., n. r. *Bel ex.* 30.—

289. **— Slee, J. C. van**, De Rijnsburger collegianten. Geschiedkund. on: derzoek. Haarlem, 1895. Av. 4 pl. d. rel. 4.50
Verhandel. Teyler's Godgel. Genootschap. N. S. XV.

290. **Schimmel, W. F.**, Geschiedkundig overzicht van het muntwezen in Nederland. Amst. 1882. 3.—

291. **Schlettwein, J. A.**, Archiv für den Menschen und Bürger in allen Verhältnissen oder Sammlung von Abhandlungen, Vorschlägen, Plänen...., welche das Wohl und Wehe der Menschheit und der Staaten angehen. Lpz. 1780–84. 8 vols. — Neues Archiv für den Menschen und Bürger. Lpz. 1785–88. 5 vols. — Tog. 13 vols. Hfcalf. 95.—
Complete collection. Extremely scarce.

292. **Sciences maudites, Les.** Sous la dir. de Jollivet-Castelot, P. Ferniat e.a. Paris, 1900. Av. pl. p. L. Galand, P. Cirou, e.a. *Imprimé sur papier rouge, bleu et blanc.* 3.—
Contient: L'occultisme contempor. en France. — L'astrologie. — La Cabbale. — L'alchimie. — Homunculus. — Clairvoyance. — etc.

293. **Sea-serpent. — Oudemans Jz., A. C.**, The great sea-serpent. Histor. and critical treatise. W. the reports of 187 appearances, etc. Leiden, 1892. W. 82 ill. cloth. 12.—

294. **Shorthand. — Congres, Internat.**, voor stenografie. Amsterdam 1934. Handelingen. Amsterdam, 1935. Av. ill. 3.—

295. **— — 2e, de sténographie.** Paris 1889. Compte rendu. Paris, 1890. cart. 4.50

296. **— Congreso internac., 10°**, de estenografia. Madrid 1912. Actas. Madrid, 1914. W. some portr. and pl. 6.—

297. **— Gabelsberger, F. X.**, Anleitung zur deutschen Redezeichen-kunst oder Stenographie. München, 1834. Av. titre calligraphié. gr. in-8vo. d. rel. 45.—
Premier ouvrage fondamental sur la sténographie. La 1re partie traite de

i

l'histoire et de la théorie de la sténographie, la 2e de la pratique; celle-c contient des exemples.

298. **Shorthand.** — **Handschrift, Die Kasseler,** der Tironischen Noten samt Ergänzungen aus der Wolfenbüttler Handschrift. Hrsg. von F. Ruess. Lpz. 1914. Av. 150 pl. gr. in-4to.   20.—
„Tironische Noten, die Stenographie der alten Römer, benannt nach ihrem Erfinder M. T. Tiro."

299. — **Verhandelingen** over de stenographie, op de Nederlandsche taal toegepast. Brussel, 1829. Av. 8 pl. lithogr. 4to. cart. *Rare*. 30.—
Contient: **H. Somerhausen,** Proeve eener Nederlandsche stenographie. — **J. Bossaert,** Ontwerp eener Nederlandsche stenographie.

300. **Sillevis Smitt, P. A. E.,** De organisatie van de Christelijke kerk in den apostoliscĥen tijd. Rott. 1910.   2.50

301. **Sloet, L. A. J. W.,** De dieren in het Germaansche volksgeloof en volksgebruik. 's-Grav. 1887.   3.50

302. **Snow, A. H.,** The question of aborigines in the law and practice of nations. Incl. a collection of authorities and documents. N. York, 1921. toile. (7.50)   4.50

303. **Sonnenfels, J. von,** Ueber die Liebe des Vaterlandes. Wien, 1771. pet. in-8vo. d. rel.   5.—

304. **South and Central America.** — **Annaes da Biblioteca Nacional do Rio de Janeiro.** Rio de Janeiro, 1876–1936. Vol. 1–50. 50 vols. 425.—

305. — **Belly, F.,** Percement de l'Isthme de Panama par le canal de Nicaragua. Paris, 1858. W. 3 maps.   7.50

306. — **Calmon, P.,** Historia social do Brasil. 2a ed. Rio de Janeiro, 1937. 2 vols.   7.50
Bibl. pedagocica brasileira. Vol. 40, 83.

307. — **Cuervo Marquez, L.,** Independencia de las colonias Hispano-Americanas. Participacion de la Gran Bretaña y de los Estados Unidos. Legion Britanica. Bogota, 1938. 2 vols. W. numer. portraits.   15.—

308. — **Documentos historicos.** Publ. do Archivo Nacional e da Bibliotheca Nacional. Rio de Janeiro, 1928–38. Vols. 1—40. 40 vols. 225.—

309. — **Gonzaga da Silva Leme, L.,** Genealogia Paulistana. S. Paulo, 1903–05. 9 vols.   25.—
Généalogie de familles patriciennes de S. Paulo (Brésil).

310. — **Hartman, C. V.,** Archaeological researches in Costa Rica. Stockholm, 1901. W. map, 87 coloured and plain pl. and 486 ill. roy. 4to. cloth.   75.—
Out of print.

311. — **Informação geral** da capitania de Pernambuco (1749). Rio de Janeiro, 1908.   6.—
Reprint of 380 pp. of vol. 28 of „Annaes da Bibliotheca Nacional". Library stamp on titlepage.

312. — **Lerdo de Tejada, M. M.,** Apuntes histor. de la heroica ciudad de Vera Cruz, preced. de una noticia de los descubrimientos hechos en las islas y en el continente americano. Mexico, 1850–58. 3 vols. W. 2 portr. and 8 maps and pl. Hfcalf.   75.—

313. — **Maximilian zu Wied-Neuwied,** Reise nach Brasiliën, 1815–1817. Frankfurt, 1820. 2 vols. W. 19 pl. roy. 4to and 2 vols. w. 3 maps and 22 pl. (5 coloured). Tog. 2 vols. fol. Hfcalf.   145.—

Prices are in Dutch guilders

314. **South and Central America. — Muñoz, G. Otero,** Semblanzas Colombianas. Bogota, 1938. 2 vols. W. 31 portr.   12.—
 I. Cronistas primitivos. — Literatos de la Revolucion. — Escritores de la Gran Colombia. — II. Prosistas y poetas de la Nueva Granada. Biblioteca de historia nacional. Vol. 55, 56.

315. **— Revista do Instituto historico e geographico do Brasil.** Rio de Janeiro, 1839-1936. Vol. 1-113. Bound in 114 vols. cloth.   1250.—
 Complete set of the principal historical journal of Brasil. The vols. 1, 2, 5 and 7 in 3d edition, the vols. 6, 8—19 and 21 in 2d edition. (There is no difference between the editions).

316. **— Reyles, C.,** Historia sintetica de la literatura Uruguaya, 1830-1930. Montevideo, 1931. 3 vols. cloth.   32.50

317. **— Rivas, R.,** Los fundadores de Bogota. 2a ed. aument. Bogota, 1938. Vol. I.   7.50
 Biblioteca de historia nacional. Vol. 57.

318. **— Rosa, M. de la,** Los conquistadores de los chibchas. Bogota, 1935.   1.—

319. **— Spix, J. B. von, e C. F. P. von Martius,** Viagem pelo Brasil. Trad. brasileira. Rio de Janeiro, 1938. 3 vols. of text and 1 vol. of 45 maps and pl.   15.—
 Traduction portugaise de l'ouvrage célèbre: Reise in Brasilien auf Befehl S. M. Maximilian Joseph I. München, 1823-31. 3 vol. de texte in-4to et 1 Atlas in-fol.

320. **— Unanúe, H.,** Observac. s. el clima de Lima y sus influencias en los seres organizados en especial el hombre. 2a ed. Madrid, 1815. Av. 2 tabl. météorol. des années 1799 et 1800. 4to. veau espagnol.   24.—

321. **— Wätjen, H.,** O dominio colonial Hollandez no Brasil. Rio de Janeiro, 1938.   4.50
 Biblioteca pedagogica Brasileira. Vol. 123.
 — See also: **Surinam.**

322. **Storm de Grave, A. J. P.,** Elcheira, Bertha, Laura, Seraphina, Blanca, Irena, Emma. Zevental verhalen. Amst. 1850. Av. titre lith. av. vign. cart.   2.50

323. **Surinam. — Benoit, P. J.,** Voyage à Surinam. Description des possessions néerland. dans la Guyane; 100 dessins pris sur nature par l'auteur, lithograph. par Madou et Lauters. Brux. 1839. W. 49 pl. large fol. hfcloth.   20.—

324. **— Berkel, A. van,** Amerikaansche voyagien, behelz. een reis naar Rio de Berbice, gelegen op het vaste land van Guiana, aan de Wildekust van America, mitsg. een andere na de colonie van Suriname.... met alle de bijzonderheden noopende de zeden, gewoonten, en levenswijs der inboorlingen, boomen, aardgewassen, waaren en koopmanschappen, en andere aanmerkelijke zaaken. Amst., J. ten Hoorn, 1695. W. front. and 2 folding pl., engraved by Luyken. 4to. hfcalf. (Mod. binding).   150.—
 A very interesting voyage for the knowledge of the manners and customs of the Indians of Guyana. One of the best Dutch books on the subject.

325. **— Beschrijvinge** van Guiana; des selfs cituatie, gesontheyt, vruchtbaerheyt ende ongemeene profijten en voordeelen boven andere landen. Hoorn, S. J. Kortingh, 1676. 4to. boards.   100.-
 Knuttel, nr. 11390. Especially treating of the bad condition of commerce and navigation of Guyana, means of improvement, etc.
 One of the scarcest items on Guyana.

326. **Surinam.** — **Beschrijvinge, Pertinente,** van Guiana. Gelegen aen de vaste kust van America. Waer in .... het aenmerckelijkste dat in en omtrent het landt van Guiana valt, als de limiten, het klimaet en de stoffen der landen, de mineralen...., vruchten, dieren.... Als oock de conditien van de Staten van Hollandt, voor die gene die nae Guiana begeeren te varen. Amst., J. C. ten Hoorn, 1676. W. woodcut on titlepage and a map. 4to. vellum.   225.—
,,Accurate description of Guiana situated on the coast of America. With addition of the profits of the share holders and the conditions for those who wish to sail for Guiana".
A very interesting item containing the conditions granted by the States of Holland to all willing to found a colony on the coast of America.

327. — **(Beyer, E.),** Suriname in deszelfs tegenwoord. toestand. (U. h. Hgd). Amst. 1823. W. vign. (view on Paramaribo) on titlepage.
7.50

328. — **Breugel, C. van,** Dagverhaal van eene reis naar Paramaribo en verdere omstreken in Suriname. Amst. 1842.   5.—

329. — **Bueno de Mesquita, J. A.,** en **F. Oudschans Dentz,** Geschiedkundige tijdtafel van Suriname (1613-1924). 's-Grav. 1925. 2.—

330. — **Coll, C. van,** Gegevens over land en volk van Suriname. 's-Grav. 1903. *Extr.*   3.—

331. — **Conduct, The,** of the Dutch, rel. to their breach of treaties with England. Partic. of the articles of capitulation, for the surrender of Surinam, in 1667; and their oppressions committed upon the English subjects in that colony. With full account of the case of Jeronimy Clifford, ...The parliamentary and other proceedings upon (his) case, during the reigns of King William and Queen Anne.... London, 1760. calf. (Binding broken).   48.—

332. — **Douglas, Ch.,** Aanteekeningen over allerlei in verband met de bevolking en den landbouw in het verleden en heden van Suriname. Paramaribo, 1928.   1.50

333. — **Eeuw, Een halve,** in Suriname, 1866-1916. Ter herinnering aan het gouden jubilé van de aankomst der eerste Redemptoristen in de missie van Suriname. 's-Hert. 1916. Av. carte, portr. et ill. 4to-obl.   5.—

334. — **Eilerts de Haan, J. G. J. W.,** Verslag van de expeditie naar de Suriname-rivier (30 Juni—20 Nov. 1908). Leiden, 1910. Av. cartes, pl. et ill. *T. à p.*   3.50

335. — **Encyclopaedie van Nederlandsch West-Indië.** Onder red. van H. D. Benjamins en J. F. Snelleman. 's-Grav. 1916. W. 3 maps. cloth.   20.—

336. — **Gids, De West-Indische.** Onder red. van H. D. Benjamins, J. Boeke, J. F. Snelleman, e.a. 's-Grav. 1919-38. Year 1-20. W. index. 21 vols. cloth. (384.—)   218.—
Excellent periodical and the only one treating exclusively of the Netherlands West-Indies, its history, politics, population, economical condition, etc.

337. — **Grol, G. J. van,** De grondpolitiek in het Westindische domein der Generaliteit. I. Algemeen histor. inleiding. 's-Grav. 1934. cloth.
2.—

Prices are in Dutch guilders

338. **Surinam.** — (**Heeckeren, E. L. van**), Aanteekeningen betr. de kolonie Suriname. Arnhem, 1826. Av. 1 (sur 2) cartes. 3.—
Traite e.a. des finances et de l'enseignement.

339. — **H(erlein), J. D.,** Beschryvinge van de volk-plantinge Zuriname, vertonende de opkomst dier zelver colonie, de aanbouw en bewerkinge der zuiker-plantagien, den aard der Indianen, als ook de slaafsche Afrikaansche mooren.... de bosch-grond, water- en pluimgedierten, vrugten, gommen, olyen en de gesteltheit van de Karaïbaansche kust. Leeuwarden, M. Injema, 1718. W. front., map and pl. 4to. Hfvellum. *Very tall copy.* 60.—
Pp. 249–262: Karaïbaansch woordenboek.

340. — **Keye, O.,** Kurtzer Entwurff von Neu-Niederland und Guajana einander entgegen gesetzt, umb den Unterscheid zwischen warmen und kalten Landen herausz zu bringen, und zu weisen welche von beyden am füglichsten zu bewohnen,.... Denen Patronen, so da Colonien an zu legen gesonnen.... A. d. Holländ. d. T. R. C. S. C. S. (Th. Ritsch). Lpz. 1672. 4to. boards. 325.—
Asher no. 12. The German edition also is of great rarity.

341. — **Lans, W. H.,** De oorzaken van verval en de middelen tot herstel der Surinaamsche plantaadjen. 's-Grav. 1829. 4.—

342. — — Bijdr. tot de kennis der kolonie Suriname. 's-Grav. 1842. 2.50

343. — — Emancipatie door centralisatie. Schets van een ontwerp tot behoud van Suriname. 's-Grav. 1847. W. 6 folding plans. *Very scarce.* 10.—

344. — **Lennep Coster, G. van,** Aanteek. gedur. mijn verblijf in de West-Indiën, 1837–1840. Amst. 1842. W. 3 pl. 5.—
Paramaribo, Porto Cabello, Port au Prince.

345. — **Oudschans Dentz, F.,** en **W. R. Menkman,** West-Indië. 's-Grav. 1933. W. 3 coloured maps. (9.—) 4.50
Part of: Geschiedkundige atlas van Nederland.

346. — **Pistorius, Th.,** Beschryvinge van Zuriname, waar in de gelegenheid deezer volkplantinge, derzelver rivieren, kreeken, forten, waterwerken, deszelfs inwoonderen, de leevensmanier der slaaven, de vrugt- en andere boomen, en dieren; een berigt van het zuiker-riet, zuiker- en koffy-plantagien, moolens....; een verhaal van de moord aan van Sommelsdyk. Amst., Th. Crajenschot, 1763. W. 4 pl. 4to. Hfcalf, uncut. 35.—

347. — **Plante Fébure, J. M.,** West-Indië in het Parlement, 1897–1917. Bijdrage tot Nederlands koloniaal-politieke geschiedenis. 's-Grav. 1918. (5.40) 2.—

348. — **Poll, E. v. d.,** Brieven betr. de plantagiën Waterland, Adrichem, Palmeribo en Surinombo in Surinamen. Amst. 1833–34. 4 pièces. 5.—

349. — **Pyttersen, Tj.,** Een deel der taak van Nederland in Suriname. 's-Grav. 1927. 1.20

350. — — Europeesche kolonisatie in Suriname. Geschiedkund. schets. 's-Grav. 1896. 1.—

351. — **Raders, J. E. W. F. van, en D. L. Wolfson,** Verslag eener reis naar Demerary, Grenada en Guadeloupe in 1845. Paramaribo, 1846. Av. pl. 7.50
Rapport sur les plantations de sucre. Très rare. Ex. sans titre.

32   SURINAM—TESTAMENT

352. **Surinam. — Raders, R. F. van,** Geschiedkund. aanteek. rakende proeven van Europeesche kolonisatie in Suriname. 's-Grav. 1860. Av. carte. toile. 4.—

353. **— Roos, P. F.,** Eerstelingen van Surinaamsche mengelpoëzy. Amst. 1783. W. vign. on titlepage. calf. 20.—
All the poems are rel. to the plantations and other places in Surinam or to different events in that colony.

354. **— Surinaamsche mengelpoëzy.** Amst. 1804. 4to. Russia. 20.—
Contains i.a.: „De suikerbouw", „Wandeling naar de plantaadje Ma Retraite", „De plantaadje Berg en Daal", „America, treurzang", „Lijkzang voor George Washington, overleden in N.-Amerika, 18 Dec. 1799", etc.

355. **— Singi vo emancipatie.** Eerste Juli 1873. Paramaribo, 1873. 2.50
Sept chansons en hollandais des nègres du Surinam.

356. **— Sterre, D. vander,** Zeer aanmerkelijke reysen gedaan door J. E. Reining meest in de West-Indien en ook in veel andere deelen des werelds, waar in verhandelt werd hetgeen hem van zyn kintsche jaren avontuurlyk ter zee en te land tot zijn 49ste jaar is voorgevallen soo tegens de Wilden als voor en tegens de Spanjaarden, Engelschen, Franse, Portugese en meer andere natien. Amst., J. ten Hoorn, 1691. W. front. and 6 pl. by Luyken. 4to. hfcalf. 175.—
Tiele, no. 1052. The editor (v. d. Sterre) of these very adventurous voyages was physician on Curaçao. The work contains a narrative of Reinings adventures in Surinam, on St Domingo, Martinique, Aruba, Jamaica, Panama, Vera Cruz, etc.
Extremely scarce.

357. **— Stuger, J. L.,** Voorheen en thans. Over Coronie's verleden en toekomst, 1808–1900. Paramaribo, 1900. 5.—

358. **— Sypesteyn, C. A. van,** Beschrijving van Suriname. Histor., geograph. en statist. overzigt. 's-Grav. 1854. W. map. 4.—

359. **— — Mr. J. J.** Mauricius, Gouverneur-Generaal van Suriname, 1742-1751. 's-Grav. 1858. W. portr. and autogr. in facs. boards. 3.50

360. **— Teenstra, M. D.,** De negerslaven in de kolonie Suriname. Dordrecht, 1842. cart. orig. 4.—
Les pp. 1—79 traitent de la colonie, de son agriculture et de sa population (libre), les pp. 177—308 de l'incendie de Paramaribo (1832) et de l'exécution de 3 nègres, les pp. 311—380 contiennent une bibliographie raisonnée.
Sans la carte et la pl.

361. **— Toestand, De economische,** van Suriname in 1922. Paramaribo, 1923. 0.75

362. **— Vestiging, De,** van de Nederlandsche kolonisten in Suriname herdacht, 1845-1870. Paramaribo, 1921. Av. 1 pl. 1.—

363. **— Wolbers, J.,** Geschiedenis van Suriname. Amst. 1861. With portr. and facs. hfcalf. *Out of print.* 15.—
A part of the inner margins slightly stained.

364. **Testament, Oude.** Opnieuw uit den grondtekst overgezet m. inleid. en aanteek. d. A. Kuenen, I. Hooykaas, W. H. Kosters en H. Oort. Leiden, 1899, 1901. 2 vol. toile. (29.—) 18.—

Prices are in Dutch guilders